D0045944

Annus Mirabilis

ALBERT EINSTEIN DURING THE ANNUS MIRABILIS.

Annus
Mirabilis

1905,
ALBERT EINSTEIN,
AND
THE THEORY OF RELATIVITY

John Gribbin and Mary Gribbin

Chamberlain Bros.
a member of Penguin Group (USA) Inc.
New York

We are grateful to the Alfred C. Munger Foundation for assistance with our travel and research expenses, and thank the University of Sussex for continuing to provide a base for us.

CHAMBERLAIN BROS.
Published by the Penguin Group
Penguin Group (USA) Inc., 375 Hudson Street, New York,
New York 10014, USA
Penguin Group (Canada), 10 Alcorn Avenue, Toronto, Ontario M4V 3B2, Canada
(a division of Pearson Penguin Canada Inc.)
Penguin Books Ltd, 80 Strand, London WC2R 0RL, England
Penguin Ireland, 25 St Stephen's Green, Dublin 2, Ireland
(a division of Penguin Books Ltd)
Penguin Group (Australia), 250 Camberwell Road, Camberwell,
Victoria 3124, Australia (a division of Pearson Australia Group Pty Ltd)
Penguin Books India Pvt Ltd, 11 Community Centre, Panchsheel Park,
New Delhi—110 017, India
Penguin Group (NZ), Cnr Airborne and Rosedale Roads, Albany, Auckland 1310,
New Zealand (a division of Pearson New Zealand Ltd)
Penguin Books (South Africa) (Pty) Ltd, 24 Sturdee Avenue, Rosebank,
Johannesburg 2196, South Africa

Penguin Books Ltd, Registered Offices: 80 Strand, London WC2R 0RL, England

Copyright © 2005 by John and Mary Gribbin
(This excludes Appendix B.)

All rights reserved. No part of this book may be reproduced, scanned, or distributed in any printed or electronic form without permission. Please do not participate in or encourage piracy of copyrighted materials in violation of the author's rights. Purchase only authorized editions. Published simultaneously in Canada

An application has been submitted to register this book
with the Library of Congress.

ISBN 1-59609-144-4

Printed in the United States of America

1   3   5   7   9   10   8   6   4   2

Book design by Jaime Putorti

Photo on p. ii Bettmann/CORBIS; Photo on p. vi CORBIS SYGMA;
Photo on p. x Bettmann/CORBIS; Photo on p. 42 Underwood & Underwood/CORBIS;
Photo on p. 106 Bettmann/CORBIS

CONTENTS

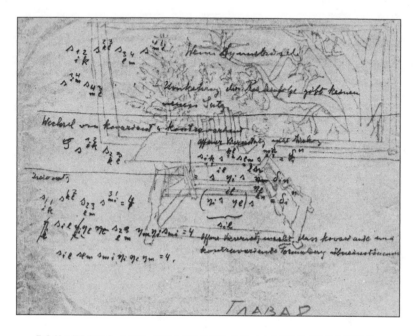

EQUATIONS AND DRAWINGS BY ALBERT EINSTEIN.

INTRODUCTION

In the history of science, the term "annus mirabilis" (literally, "miraculous year") usually refers to the months in 1665 and 1666 when Isaac Newton made his great discoveries in optics, mathematics, and his understanding of the nature of gravity. But in 1905, Albert Einstein had his own annus mirabilis, when he published four scientific papers that had a profound influence on the science of the twentieth century. Everybody knows Einstein's name, and an equation from one of those papers, $E = mc^2$, is the most famous equation in all of science. But who was the man who made those breakthroughs, and why was his work so important? From our perspective exactly a hundred years on from Einstein's annus mirabilis, we hope to be able to answer those questions for you.

As ever, we are grateful to the University of Sussex and the Alfred C. Munger Foundation for providing, respectively, a base for us and a contribution toward our travel and other expenses.

John Gribbin
Mary Gribbin
December 2004

ALBERT EINSTEIN WITH HIS SECOND WIFE,
ELSA EINSTEIN.

CHAPTER ONE

The First Twenty-five Years

In 1905, Albert Einstein produced the most important package of ideas from any scientist since Isaac Newton. The iconic image we have of Einstein is the white-haired genius, a wise and fatherly guru, a cross between God and Harpo Marx. But in 1905 Einstein was a handsome, dark-haired young man (he celebrated his twenty-sixth birthday on March 14 that year), previously something of a ladies' man but recently married, with a baby son. He didn't even have a PhD at the time he published the scientific papers that made him famous.

What made Newton and Einstein so special was that they didn't just have one brilliant idea (like, say, Charles Darwin with his theory of natural selection) but a whole variety of brilliant

ideas, within a few months of each other. There were other similarities between the two great men. In 1666, Newton celebrated
his twenty-fourth birthday, and although he had already obtained his degree from the University of Cambridge, for much of
1665 and 1666 he had been unable to take up a Fellowship at
Trinity College because the university had been closed by an
outbreak of plague. So he had been working in isolation at the
family home in Lincolnshire. In 1905, the twenty-six-year-old
Einstein had already graduated from the Swiss Federal Institute
of Technology, but had been unable to obtain a post at any university. So he had settled for a junior post at the patent office in
Bern, working in isolation on scientific topics at home in his
spare time—and also, as he later admitted, during office hours.

Especially in the theoretical sciences and mathematics, it is
often true that people do their best work in their twenties, even
if that work never matches the achievements of a Newton or an
Einstein. But there the similarities between the two geniuses
stop. Newton was a loner by choice, who made few friends and
never married; although most of his great work was done in
1665–1666, it was only published later, at different times, in response to pressure from colleagues who became aware of what
he had achieved. Einstein was a gregarious family man, eager to
get a foothold in the academic world, who knew the importance of advertising his discoveries and published them as soon
as he could. It was his only chance of getting out of the patent
office and into a university post. But what was the man later regarded as the greatest genius of the twentieth century doing
working in a patent office anyway?

Albert had been born in Ulm, in Germany, in 1879. In the summer of the following year, however, the family moved to Munich, in the south of Germany, where Albert's father, Hermann, went into partnership with his younger brother Jakob in the booming electrical industry. Jakob had a degree from the Stuttgart Polytechnic Institute, and provided the expert know-how for the business; the money to set them up came from the family of Albert's mother, Pauline. Hermann's role was on the administrative side, running the business. Jakob and his wife, Ida, shared a pleasant house on the outskirts of Munich with Hermann, Pauline, and little Albert.

Far from showing any signs of precocious genius, little Albert was so slow to learn to speak that his parents feared that there might be something wrong with him. It wasn't until well after his second birthday that he began talking, but when he did he used proper sentences from the start, quietly working the words out in his head and whispering them to himself before speaking out loud. In November 1881, his sister, Maja, was born, and she later recalled his response, reported to her by her mother when she was old enough to understand. It seems that Albert had been told he would soon have something new to play with, and was expecting a toy; on being introduced to his baby sister, he asked, "Where are the wheels?"[1] Although brother and sister developed an affectionate bond, young Albert was prone to occasional outbursts of violent temper, when he would throw

---

[1] Maja's reminiscences are preserved in the Einstein archive at Princeton University.

things at the nearest person—all too often, Maja. Family legend tells of the time he hit her over the head with a garden hoe, and how at the age of five he chased his first violin teacher out of the house, throwing a chair after her. But the music-loving Pauline was tough and strong-minded enough to ensure that he carried on with the violin lessons whether he liked it or not, eventually instilling in him a love of music that provided a lifetime release from the strains of his scientific work. And he had talent—much later, when he was a sixteen-year-old high school student in Switzerland, a school inspector would single him out for praise, reporting that "one student, [named] Einstein, actually sparkled [in] his emotional performance of an adagio from a Beethoven sonata."[2]

The expression "tough love" could have been invented to describe Pauline's attitude to her children. When he was only four, Albert was given a guided tour of the neighborhood by his parents, and from then on was not only allowed but encouraged to go out alone and find his way through the streets— although, unknown to the boy, they kept a distant eye on him on his first few solo expeditions. One of the first regular journeys he had to make was to school. The Einsteins were secular Jews, and were unconcerned that the nearest school was Catholic, so that is where Albert received his first formal education, very much in the old-fashioned tradition of learning by rote and with strict discipline enforced by corporal punishment. Albert's disenchantment with the school was strengthened

---

[2] *Collected Papers.*

when he was still only five years old, and ill in bed. His father gave the boy a magnetic compass to relieve his boredom, and Albert became intrigued by the way in which the needle always tried to point to the north, no matter how he twisted and turned the instrument. He was fascinated by the idea of an invisible force that kept a grip on the compass needle, and baffled that none of his teachers at the school had shown him anything half as interesting. This helped to instill an early conviction that he was much better off working things out for himself than working within the system.

Albert's stubborn insistence on finding his own way in the world led to a curious development during his years at the Catholic school. At that time, the statutes of the city of Munich required all students to receive some religious education, and although Hermann and Pauline didn't mind Albert attending a Catholic school, they drew the line at having him indoctrinated in the Catholic faith. So, to meet their obligations they got a relative to teach Albert about the Jewish faith—as they thought, just as a matter of form. To everyone's surprise, Albert lapped it all up and became something of a religious fanatic, observing the Jewish rituals that his parents had abandoned, refusing to eat pork, and making up hymns that he would sing to himself as he walked to school in the morning. This religious phase lasted until Albert was about twelve, and had been a student at the local high school (*Gymnasium*) for two years. His loss of faith was a direct result of his discovery of science; but that discovery owed nothing to the high school, and everything to a young medical student named Max Talmey.

Although Hermann and Pauline Einstein did not follow all the religious traditions of their nominal faith, there was one Jewish custom that they kept. At that time, there was a tradition among middle-class Jewish families of helping young students who might be struggling to make ends meet, and the Einsteins got into the habit of inviting Talmey, who came from Poland,[3] to dinner once a week. It was Talmey who introduced Albert, who was ten when they met, to the latest scientific ideas, discussing them with the boy as if he were an adult. Talmey lent books popularizing science to Albert, introduced him to algebra, and in 1891 gave him a book about geometry. Einstein later described reading this book as the single most important factor in making him a scientist. He was gripped, fascinated by the way in which mathematical logic could be used to start from simple premises and construct truths, such as the Pythagorean theorem, that are *absolutely* true. Within a year, he had worked his own way through the entire mathematics syllabus of the high school. Which rather left the school, as he saw it, as a pointless waste of time. He had also lost his faith in God, and now, as a young teenager, saw religion as part of a deception played by the State in order to manipulate its people, especially the young.

Albert had made a good start to his time at the Luitpold-Gymnasium, finding it easy to keep up with his peers; but although this was one of the best schools of its kind, a combination of the rigid educational system, his loss of religious

---

[3] His name is sometimes translated as "Talmud."

faith, and his discovery that he could leap ahead of the curriculum by working on his own led him to neglect his classwork. He still got good grades in mathematics, but couldn't see the point of subjects like Classical Greek and gained a reputation as an impudent troublemaker.

The situation at school came to a head in 1894, when the family business went bust. The firm had done reasonably well in the 1880s, like many other small businesses taking advantage of the booming demand for electrical equipment such as dynamos, lighting systems, and telephones. But in the classic pattern of progress following the invention of a new technology, those small businesses were now being swallowed up by large firms, or going under in the face of competition from the giants, such as Siemens and AEG. It was a lack of capital, as much as anything else, that saw the Einsteins lose out in this competition. But before matters came to a head they had gained a reputation in southern Germany and northern Italy. Italy had not gone so far down the road toward domination of the electrical industry by one or two firms, and with the encouragement of one of their Italian colleagues, Hermann, Jakob, and their wives decided to move to Italy and start again. There was one snag. Albert, who was now fifteen, had three years left to complete at the Gymnasium, which would ensure his admission to a good university. A family decision was made to leave him behind in Munich, staying in a boardinghouse but with a distant relative keeping an eye on him.

The result ought to have been predictable. Albert was lonely and miserable in Munich, and without family life to fall back on the school seemed unbearable. There was another worry. Albert had always hated the militaristic aspect of German society at the time, and as a young boy, scared by the sight and sound of marching troops, had once begged his parents to promise that he would never have to be a soldier. But the law then required every German male to undergo a period of military duty. The only escape was to leave the country, and renounce his German citizenship, before his seventeenth birthday; if he left later, he would be regarded as a deserter.

There is some confusion about how exactly Einstein engineered his escape from Munich, but this is the most likely pattern of events. He was certainly depressed, and managed to persuade his doctor (the family doctor before the family moved to Italy) to provide a certificate stating that he should rejoin the family for health reasons. He also persuaded his math teacher to provide a letter stating that Einstein had already mastered the syllabus, and there was nothing more he could teach the young man. Armed with these documents, Einstein approached the principal of the school and told him that he was leaving; the principal's response, we are told, was that Einstein was being expelled anyway, for being a disruptive influence. The likelihood is that Einstein had carefully cultivated his role as a disruptive influence in order to ensure that he would not be asked—let alone ordered—to stay on; but whether he jumped or was pushed, there is no doubt that in the spring of 1896, just six months after being left in Munich to complete his education, Einstein turned up on the doorstep of his parents' new home in Milan, in northern Italy.[4] He duly renounced his German citizenship (the declaration took effect in January 1896), and swore he would never go back to that country. By then, Einstein was living in Switzerland, and duly set in motion the slow process of obtaining Swiss citizenship; but in those relatively free and easy days in Europe the lack of a passport did not prevent him from traveling as he pleased, in particular between Switzerland and Italy. On official forms, under "nationality" he simply described himself as "the son of German parents."

In the summer of 1895, Einstein was sixteen, in Italy, with no responsibilities (and no prospects). Although he did some work for the family business, and had vague notions of becoming a teacher of philosophy, for several months he mostly just enjoyed himself, touring the art centers of Italy, visiting the Alps, and falling in love with the culture. When pressed by his father to settle down and give some thought for the future, Einstein assured Hermann that in the autumn he would take the entrance examination for the Swiss Federal Polytechnic in Zurich (often known by the initials of its German name, as the ETH). The ETH was not a great university in the mold of Heidelberg or Berlin, but a new kind of institution devoted primarily to the education of would-be teachers and engineers. The cocksure Einstein was certain he would be able to walk into an

---

[4] Shortly after Albert arrived, the family moved to the smaller town of Pavia, near Milan.

establishment with such relatively modest academic standards, and it came as a rude shock when he failed the exam (his later claim to have failed deliberately, in order to avoid being pushed into a profession by his father, should be taken with a large pinch of salt; if that were the case, why would he have gone back and embarked on the same course a year later?).

In fact, Einstein was lucky to be allowed to take the exam. The normal age for this was eighteen, and applicants were expected to have a high school diploma; Einstein was still six months short of his seventeenth birthday, and had left high school under a cloud. But the Director of the ETH, Albin Herzog, recognized Einstein's potential, and offered him a lifeline. He suggested that Einstein should enroll in a Swiss secondary school in the town of Aarau, a little way outside Zurich, and do some catching up before taking the entrance exam again in 1896. There, he could lodge in the home of Jost Winteler, a teacher at the school, and live as one of the family.

The domestic arrangements suited Einstein down to the ground. Winteler was not one of Einstein's tutors, so school did not intrude too much on daily life. One of Einstein's cousins, Robert Koch, was a student at the school and lodging next door. And there was plenty of family life—the Wintelers had three daughters and four sons, plus a couple of other paying guests, and Einstein was soon one of the family. A classmate during this year in Aarau later described the Einstein of 1895–1896 to his biographer Carl Seelig as "sure of himself, his gray felt hat pushed back on his thick, black hair [striding] energetically up and down . . . unhampered by convention, his

attitude toward the world was that of the laughing philosopher, and his witty mockery pitilessly lashed any conceit or
pose."

Mature beyond his years, Einstein clearly made a big impact
on his companions—none more so than Marie Winteler, the
eighteen-year-old daughter of the house, who had just completed her course as a trainee teacher and was living with her
parents while waiting to start her first job. In spite of the age
difference, an enormous gulf for most teenagers, the two fell in
love. Both families seem to have been happy with the development, which blossomed into something like an unofficial en-
gagement and persisted after Einstein returned to Pavia and
then moved on to Zurich. But in the spring of 1897, as his own
horizons broadened and he made new friends in the city, he
broke off the relationship. It took some time to convince Marie
that he meant it, but in the end everything was settled amica-
bly, and the Einsteins and the Wintelers remained good friends—
so much so that Einstein's sister Maja later married Marie's
brother Paul. Anna, another of the Winteler siblings, married
Michele Besso, Einstein's best friend in Zurich and for long after
his student days were over. In a letter[5] written to Besso's wife and
son after he died in 1955, Einstein said, "What I admired most
about Michele was the fact that he was able to live so many years
with one woman, not only in peace but also in constant unity,
something I have lamentably failed at twice." Perhaps signifi-
cantly, Marie did not marry until 1911.

---

[5] Quoted by, for example, Dennis Overbye.

Alongside his happy relationship with the Winteler family, an active social life, music, and his first experience of love, Einstein did enough work at school (where he had been allowed to join the final year group, with classmates a year older than himself) to achieve high grades in all his subjects except French, which he always struggled with. In the Swiss system, examination papers were marked on a grading scale from 1 to 6, with 6 being the top mark. Einstein's average of 5½ was the best in his year, from the youngest pupil in the class. Although this was brought down by his French paper, which was rather generously graded 3–4, the essay he wrote (in French) for the examination is the most interesting of the papers (which survive and can be found in *The Collected Papers of Albert Einstein*) because the set subject of the essay was *Mes Projets d'avenir*.[6] Ignoring the terrible French, the essay gives us a glimpse into Einstein's personal ambitions at the time, which seem remarkably limited in the light of what he would achieve:

> If I am lucky enough to pass my examinations, I will attend the Polytechnic in Zurich. I will stay there four years to study mathematics and physics. My idea is to become a teacher in these fields of natural science and I will choose the theoretical part of these sciences.

Einstein goes on to say that this ambition is partly based on the fact that he lacks any "practical talent," but also because

---

[6] *My Plans for the Future.*

"there is a certain independence in the profession of science that greatly appeals to me." That certainly chimes with the way his life would develop. But there was already a hint of what was to come in some of the ideas Einstein did not commit to paper in 1896. He later recalled that while still a schoolboy in Aarau, he puzzled over the idea that if you could run at the speed of light, you would see a light wave standing still alongside you, frozen in time, as it were; but the laws of physics said that such a "time-independent wave-field" could not exist.[7] It would be nine years before he found the solution to that puzzle.

After duly passing his entrance examination for the ETH early in the summer of 1896, Einstein spent some time with his family in Italy, where the electrical business was once again in crisis. Jakob left to work for another company, and ended up living comfortably in Vienna as the manager of a firm of instrument makers; Hermann tried to make another fresh start in Milan. Albert was sufficiently concerned about the prospects for this latest venture that he tried unsuccessfully to dissuade his relations from pouring more money into the scheme; but in October he had to put these family difficulties behind him as he returned to Zurich to enroll for his course.

In spite of the year spent in Aarau, Einstein was still six months short of the official age for admission, eighteen, and one of the youngest students ever admitted to the ETH. The Poly, as it was known locally, wasn't that old itself, having been founded in 1855 as the first university-level academic institution

---

[7] See Seelig.

in Switzerland (the Swiss Confederation only came into being in 1848). Since then, three universities had been established in Switzerland—in Basel, Zurich, and Geneva. Unlike the Poly, which was a Federal Swiss government establishment, the universities were run by their respective cantons. Unlike the universities, the Poly could not award doctoral degrees; but in 1911 it was given full university status, including the right to award doctorates. That hasn't stopped it being known as the Poly right down to the present day.

At the end of the nineteenth century, there were just under a thousand students at the ETH. But, as its name implies, the ETH was primarily devoted to the education of engineers, not theoretical physicists, and there were just five students, including Einstein, taking the science course in his year at the ETH. These included Marcel Grossmann, a model student who attended all the lectures and took detailed notes which he kept carefully for revision. Grossmann became a firm friend of Einstein, and in the long run those beautifully written notes would prove even more important to him than to Grossmann. The group also included a lone woman, Mileva Maric, a Serbian, from what was then part of the Hungarian Empire. Dark and slightly exotic, Mileva was almost four years older than Einstein, and had had to struggle with an unsympathetic family and unsupportive school system at home to make it to university in Switzerland, the only German-speaking country where women were admitted to university at that time. Indeed, Mileva was only the fifth woman to be admitted to study physics at the ETH. The other two members of the class were Jakob Ehrat, a hardworking but unspectacular student, and Louis Kollros; both, like Grossmann, were Swiss.

Like many students, Einstein enjoyed the freedom of university life to the full, and didn't worry too much about the academic side until the examinations loomed. He didn't reckon anyone could teach him mathematics better than he could learn on his own with the aid of books, and he seldom attended the lectures, leading one of his professors, Hermann Minkowski, later famously to describe the student Einstein as a "lazy dog," who "never bothered about mathematics at all." In fact, Minkowski was one of the few professors at the ETH that Einstein respected, and when he did attend a lecture given by Minkowski during his final semester at the ETH, Einstein remarked to Louis Kollros that "this is the first lecture on mathematical physics we have heard at the Poly."[8]

Cutting lectures gave Einstein plenty of time to indulge his passions for coffeehouse discussions with his friends setting the world to rights (including scientific discussions about the latest ideas in physics), the company of women (he always got on well with women, who were charmed by his manners, his music, and his masculine good looks), sailing on the lake (where he always took a notebook in case the wind dropped, so that he could scribble down his ideas on physics), and music (combining this with his love of the company of women, Einstein often gave recitals on the violin in the homes of ladies of his acquaintance). He lived in lodgings in Zurich, getting by

---

[8] See Overbye.

financially on an allowance of one hundred Swiss francs a month, generously provided by one of his maternal aunts, eked out by a little private tuition. Out of this, he set aside twenty francs to save toward the fee he would have to pay when he was eventually awarded Swiss citizenship. On Sundays, he took lunch with the family of Michael Fleischmann, a Zurich businessman, echoing the way the Einsteins had looked after Max Talmey in Munich.

It was music that brought Einstein and his lifelong friend Michele Besso together. Besso was six years older than Einstein, and already working as a mechanical engineer. They met at a house where Einstein was among the musicians entertaining a group of students and other people, an important, if informal, social activity in those days before television, radio, or recorded music. It was Besso who introduced Einstein to the work of Ernst Mach, an Austrian philosopher-physicist who had made important contributions to the scientific debate raging at the end of the nineteenth century concerning the reality of atoms.

From our modern perspective, it is hard to believe that only a little over a hundred years ago people were still arguing about whether or not atoms were real. But this was indeed an important debate, which would influence a great deal of Einstein's early scientific work and become a significant feature of his annus mirabilis.

Popular accounts of the history of science often tell you that the idea of atoms goes back to the time of the ancient Greeks; but this is true only up to a point. What is true is that the Greek philosopher Democritus, who lived in the fifth century B.C., did discuss the idea that everything is made of tiny, indestructible particles moving through a void (the vacuum) and interacting with one another. But this was never more than a minority view at the time, and was dismissed by most of the ancient Greek thinkers because they could not accept the idea of the void, a genuine nothingness between atoms. The idea was revived from time to time, notably by the Frenchman Pierre Gassendi in the seventeenth century, but always dismissed, for the same reason. It was only in the nineteenth century that a large group of scientists really began to take the idea of atoms seriously, and even then others argued against the idea.

The scientists who took the idea of atoms seriously found evidence supporting the idea in both chemistry and physics. In the early 1800s, John Dalton, in England, developed the idea that each element (such as hydrogen or oxygen) is made up of a different kind of atom (but with all the atoms of a particular element identical to one another), and that compound substances (such as water) are made up of molecules in which different kinds of atom join together (in this case, as H_2O). As early as 1811, jumping off from these chemical discoveries, Italian chemist and physicist Amedeo Avogadro announced his famous hypothesis, that at a given temperature and pressure equal volumes of gas contain the same number of particles (molecules or atoms), with the clear implication that there is nothing in the space between these particles. But his idea was ignored for decades, and there was no clear idea of the differences between atoms and molecules until the work of Avogadro's compatriot, Stanislao Cannizaro, in the 1850s.

By then, evidence supporting the idea of molecules was coming in from the physicists. One of the most important practical applications of science in the nineteenth century concerned the study of heat and motion (known as thermodynamics), which was directly relevant to the application of steam power during the industrial revolution. By studying the way in which heat could be generated, and how it flowed from one object to another, scientists came up with laws of thermodynamics to describe the relationship between work and energy on the scale of the kind of machinery that powered the industrial revolution—sometimes referred to as the "macroscopic" scale. Physicists such as the Scot James Clerk Maxwell, Hermann von Helmholtz in Germany, and the Austrian Ludwig Boltzmann then developed models to describe these macroscopic phenomena in terms of the accumulated effect of huge numbers of atoms and molecules bouncing around and interacting with one another like tiny, hard spheres, obeying the basic laws of mechanics discovered by Isaac Newton two hundred years earlier. This behavior of atoms and molecules at a lower level is often referred to as "microscopic" behavior; but, crucially, atoms are actually far too small to be seen by any microscope available in the nineteenth century.

The way the cumulative behavior of vast numbers of atoms and molecules interacting on the microscopic scale combines to produce measurable effects on the macroscopic scale is called statistical mechanics. For the particular case of gas trapped in a box, this approach proved an excellent way to explain how the pressure and temperature of the gas change as the box is made smaller or larger and the speed of the molecules and atoms changes; this is known as the kinetic theory, since it is all about movement.

All of these ideas were in the air in the 1890s, and formed the subject of many conversations between Einstein, Grossmann, Besso, and their friends, wreathed in tobacco smoke as they lingered over coffee in some Zurich café, or while striding through the countryside on extended walks. The problem was that although ideas like statistical mechanics and the kinetic theory worked at a practical level to provide a mathematical description of what was going on, nobody had seen atoms—more to the point, given the technology of the time it was physically impossible to see atoms. This left the door open for philosophers such as Ernst Mach to argue that the atomic hypothesis was no more than a hypothesis, what is known as a heuristic device, meaning just because things in the macroscopic world behave *as if* they were made of atoms that doesn't prove that they *are* made of atoms. Mach regarded atoms as no more than a convenient fiction, which provided a basis for physicists to make calculations; anything that could not be detected by the human senses, he argued, was not the proper subject of scientific debate.

Einstein disagreed, and argued the case for atoms with his friends. He became obsessed with the idea, and determined that if no one else could prove that atoms were real, he would do it himself. As he wrote many years later in his *Autobiographical Notes,* he determined that as soon as he had graduated from the ETH he would search for evidence "which would guarantee as much as possible the existence of atoms of definite finite size." He succeeded, as we shall see in the next chapter. Just as important as the fact that he succeeded, though, is the fact that at the beginning of the twentieth century Albert Einstein, widely regarded as the greatest genius of the twentieth century, thought that the most important problem facing science was to prove the reality of atoms. That alone shows just how far we have come in the past hundred years.

At first, Einstein did well at the ETH, his brilliance enabling him to shine in his intermediate examinations even though he had not been following the curriculum as diligently as his friend Grossmann. But alongside his studies and his discussions about the latest hot topics in science he developed another passion, one which meshed with his interest in physics but would eventually prove a distraction from his scientific work. It isn't clear exactly how or when Einstein and Mileva Maric became more than just friends, but their discussions about physics seem to have developed a romantic side by the early summer of 1897. It may be no coincidence that Einstein broke off his relationship with Marie Winteler in the spring of that year; but Mileva seems to have been the first to be seriously affected by the new relationship.

The circumstantial evidence suggests that about this time Mileva began to fall in love with Einstein, and became confused about her future prospects. She had made a huge effort, as a woman, to get a place at the ETH, a foothold on a professional career in science, and seems to have become concerned at the possibility that it might all be wasted if she were to marry, settle down, and (as would then have been almost inevitable in those days) have children. In the summer of 1897, she went home to her family as normal, and kept up a correspondence with Einstein; but at the end of the summer, instead of returning to Zurich she went, without explanation, to Heidelberg. Although this was the home of one of the great German universities, it wasn't so good for women, who were patronizingly allowed to attend lectures by special permission of individual professors, but were not allowed to take a degree. Whatever her motives for this diversion, by April 1898 Mileva was back in Zurich, where Einstein promised to help her catch up with her course work (using Grossmann's notes); but, as we shall see, academically she never did make up for this lost time.

Einstein himself sailed through the "intermediate diploma examination," held in October 1898, coming top of his small class with a grade average of 5.7; Grossmann came second with 5.6, and Mileva had to postpone the exam because of the time she had spent in Heidelberg. Einstein's promised help with her catching up proved to be more of a hindrance than a help, in more ways than one.

In his third year at the ETH, Einstein developed his interest in electromagnetism, and in particular the behavior of light. Mileva was roped in, and used as a sounding board for his ideas—much more exciting than the course work—when she should have been concentrating on the curriculum. Maxwell (who, as it happens, died in the year Einstein was born, 1879) had discovered a set of equations which describe, among other things, the way electromagnetic waves move. The equations predicted that the speed of those waves would be 300,000 kilometers per second.[9] Since the speed of light was being measured accurately at around the same time that Maxwell came up with this number (in the 1860s), and the speed of light exactly matches the predicted speed of electromagnetic waves, this was seen as proof that light is a form of electromagnetic vibration.

But what was vibrating? By analogy with the way sound waves propagate as vibrations in the air, or in other substances, in the nineteenth century physicists thought that light (and other forms of electromagnetic wave) must propagate in the form of vibrations in a tenuous substance they called "the ether." The ether was assumed to fill all of space, and even somehow to fill the atmosphere of the Earth, mingling with the air, to enable light to propagate. It would, though, have to be very tenuous, since planets, and even people, seemed to be able to move through it as if it did not exist. And yet, since the speed with which waves travel depends on the stiffness of the substance they propagate in (so that sound travels faster in steel than in air), it would have to be incredibly stiff—far stiffer than steel.

By the end of the nineteenth century, unsuccessful attempts were being made to measure the speed of the Earth through the ether, by measuring the speed of light in directions along the

[9] More precisely, 299,792.458 km/second.

line of the Earth's motion through space and at right angles to its motion; but the speed of light always seemed to be the same, whichever way the experimental apparatus was pointed. Einstein seems to have either been unaware of these experiments or unimpressed by them, but he was fascinated by Maxwell's equations and their prediction of a specific speed for light. He now had another "thought experiment" to highlight the mysterious nature of these waves. If you could run at the speed of light while holding a mirror in front of your face, would you be able to see your reflection? Presumably not, since the light leaving your face to bounce off the mirror and come back into your eyes as a reflection would never be able to catch up with the mirror!

But there was something else about Maxwell's achievement that appealed to Einstein. Maxwell had worked his ideas out entirely theoretically. He had produced his equations, and predicted the speed of light without doing an experiment. He had also predicted that there must be other forms of electromagnetic waves, what we now call radio waves, and these had duly been discovered by the German Heinrich Hertz in 1888. Or rather, Hertz had not *discovered* radio waves; he had merely *detected* what Maxwell, the theorist, had "discovered." Instead of experimenters making observations which the theorists then had to try to explain, the theorist had worked it all out on his own. This was what appealed to Einstein; the idea that the power of the human mind and mathematics were alone enough to conjure up deep truths about the world, echoing the way the

Greeks had conjured up deep truths about geometry. As early as the summer of 1899, Einstein wrote to Mileva[10] that:

> I'm more and more convinced that the electrodynamics of moving bodies as it is presented today, doesn't correspond to reality, and that it will be possible to present it in a simpler way.

And he said in the same letter that it was not possible "to ascribe physical meaning" to the concept of the ether. But he seems to have been unable to take his ideas further at that time, and the discussion of electrodynamics lapsed during his fourth and final year at the ETH. Getting ahead of our story only slightly, the idea of the ether was indeed soon made redundant, and physicists now think of light and other forms of radiation in terms of "fields," rather like the pattern of lines of force that you can see when a bar magnet is placed under a sheet of paper and iron filings are gently sprinkled on top of the paper. Once they are given a push, waves propagate through these fields until they interact with something else.

Einstein's own interaction with Mileva clearly intensified during their fourth year at the ETH, and it was probably sometime during that year that they decided to get married. They had spent the summer of 1899 apart, with Einstein on holiday with his mother and sister, and Mileva at home revising for her postponed intermediate examination. But although they still had

---

[10] *Collected Papers.*

separate lodgings in Zurich on their return, Einstein seems to have spent as much time at hers as in his own, and in a letter written in October 1899 refers to his "household" with Mileva.[11] Soon, he was addressing her as Doxerl ("Dolly," as in "little doll") and she was calling him "Johonzel" ("Johnny"). In spite of the deepening of their relationship and his neglect of the formal lectures throughout his time at the ETH, by cramming hard using Grossmann's notes Einstein was able to pass his final examinations. Not gloriously. He came fourth out of the five students sitting the exam, with a grade average of 4.9. But Mileva, who lacked his genius, was unable to make up adequately for lost time; she scored only 4.0, and failed. She did spectacularly badly at math, in particular, scoring only 2.5 out of the possible 6 marks.

By now, of course, Einstein's parents knew all (or perhaps, not quite all) about his relationship with Mileva. In spite of his rebellious nature and lack of obedience to authority, their Albert had made it through the educational system to earn a diploma which ensured him a respectable job. But when he told them, in the summer of 1900, that he intended to marry "Dolly," they were aghast. In their view he was far too young (only just twenty-one), they weren't sure about the girl (who was in any case much older than him), and a man ought to achieve security (not least, a job) before he even contemplated marriage. Even Einstein's friends in Zurich seem to have been surprised that a man who was so attractive to women should have chosen this one for his permanent partner. But part of the attraction to him

---

[11] *Collected Papers.*

seems to have been the thought of sharing his life with someone with whom he could share everything, including science. He fantasized in his letters to Mileva about them both getting PhDs and working together on scientific papers that would take the world by storm. But first, she had to retake her examinations at the ETH, and he had to get a job in Zurich at the Poly itself or at the university, so that they could be together.

It was at this point that reality began to intrude on the dream. Einstein's diploma from the ETH qualified him to work as a teacher of mathematics and science in secondary schools. What he wanted was to get a job as an assistant to one of the professors, and work for a PhD—although the ETH did not at that time award doctorates, there was an arrangement whereby any graduate from the ETH could write a dissertation for submission to the University of Zurich and be considered for the award of a PhD, so this was by no means an impossible dream, at least for the top students. But Einstein was not, on paper at least, one of the top students. He seems to have been convinced that in spite of obtaining the worst pass in his year and still having a reputation as a cocky, know-it-all troublemaker, his innate ability (of which he at least was in no doubt) would be recognized and be sufficient to obtain the post he wanted. But that was not the way the system worked. Hardly surprisingly, the men who had done best in the exams received priority in consideration for what limited opportunities were available, and in any case the authorities at the ETH were happy to see the back of Einstein, who had been far from a model student. They may well, also, have known about his relationship with Mileva, which went beyond the bounds of what was regarded as decent in that time and place.

The couple again spent the summer apart, each with their respective families, and in the face of parental opposition talk of marriage ceased. Mileva studied for her resits, while Einstein applied unsuccessfully for assistantships, took a little holiday, and tried to forget his troubles by studying Boltzmann's work on thermodynamics. In the autumn, both returned to Zurich where, in spite of the lack of any job for Einstein at the ETH, they both paid the appropriate fees and registered to work in the lab there on problems involving heat and electricity (the electron itself had only been discovered in 1897, after Einstein had begun his course at the ETH), with a view to obtaining a pair of PhDs. Einstein's allowance from his aunt had stopped when he graduated, and he had to make ends meet by private tutoring; Mileva still had a small allowance from her father. In these less than promising circumstances, Einstein completed his first scientific paper, on capillary action, which was duly published in the prestigious *Annalen der Physik*. The paper concerned the nature of the forces between molecules, and this led Einstein in a direction that would, eventually, result in the award of a doctorate. But Mileva's hopes of obtaining a PhD gradually faded away.

Apart from Einstein's first scientific paper, the other highlight of what must have been a far from happy few months in Zurich came in February 1901, when Einstein at last became a Swiss citizen. Somewhat ironically, in view of his reasons for renouncing German citizenship, he then had to take a medical examination prior to a spell of compulsory military service—but failed on the grounds of flatfeet and varicose veins. So he probably would not have been forced to serve in the German army even if he had stayed in Munich.

Just a month after becoming a Swiss citizen, though, Einstein had to admit defeat in his efforts to find suitable employment in Zurich. Under pressure from his parents to get a real job (any job!), he abandoned the increasingly faint prospect of a dissertation on thermoelectricity and returned to Italy, leaving Mileva to work toward her second sitting of the final examinations (while also trying to keep her own studies of heat conduction going), and facing an uncertain future.

In Milan, Albert at least had Michele Besso, who now lived in that city with his wife and young daughter, to talk to about physics; he also came across the new work of Max Planck, a German physicist who had just discovered that some of the properties of light and other forms of radiation could best be explained if the radiation is only emitted or absorbed by objects in the form of discrete packets of energy, which became known as quanta. His key paper, laying the foundations of what became quantum physics, had only been published in 1900, but as usual Einstein was keeping up with cutting-edge research in physics.

He was also keeping up his older interests, including his interest in light and motion. His letters to Mileva are full of the usual endearments expressed by separated lovers, but they also include references to his scientific work, and one of those references in particular has led some people to see Mileva as a more important scientific influence on him than she really was. In a letter written in April 1901[12] he writes, "I'll be so happy and proud when we are together and can bring our work on relative motion to a successful conclusion." Taken out of context, you might see that as indicating that Mileva made a significant contribution to the theory of relativity. But the proper context is that she failed her exams (not least because she spent too much time discussing esoteric ideas with Einstein instead of concentrating on her proper coursework) and she was by far the weakest mathematician in her year. She was alone in Zurich, struggling (in the end, unsuccessfully) to prepare for her resits, and Einstein was trying to cheer her up by assuring her how important she was to him. In the same letter, he writes, "You are and will remain a shrine for me to which no one has access; I also know that of all people, you love me the most and understand me the best." He was clearly besotted with her, and saw the work on what became relativity theory as "ours" in the sense that "what's mine is yours." But there is not one shred of evidence that Mileva contributed anything more to the development of Einstein's great theory than her role as a scientifically literate listener on whom he could try out his ideas.

Soon after that letter was written, things began to look up for Einstein. The first good news came in a letter from his old friend Marcel Grossmann. Grossmann had mentioned Einstein's increasingly desperate search for a job to his father, who

---

[12] Jürgen Renn & Robert Schulmann, editors, *Albert Einstein, Mileva Maric: The Love Letters*.

happened to be a friend of the director of the Swiss patent office, Friedrich Haller. Haller had told the elder Grossmann that there would shortly be a vacancy at the patent office in Bern, and that Einstein should certainly apply for the post when it was advertised. Just when the post was likely to be advertised was not clear; but the very next day Einstein received another letter offering him some temporary work. Jakob Rebstein, who had formerly been an assistant at the ETH and knew both Einstein and Mileva, now had a job as a mathematics teacher in Winterthur, near Zurich, but was about to go on his spell of compulsory military service for a couple of months. Would Einstein like to act as his locum tenens?

The job would not only help to fill the gap while waiting for the patent office post to be advertised, but meant that he would be near Mileva. To celebrate the good news, the couple met up at Lake Como for a short holiday before he took over from Rebstein. In a letter to a friend,[13] Mileva described how they rode through the snow-covered countryside in a horse-drawn sleigh:

> We drove one moment through long galleries and the next on the open road, where, all the way to the remotest distance, our eyes could see nothing but snow and more snow, so that at times I shuddered at this cold white infinity and firmly kept my arm round my sweetheart under the coats and blankets which covered us . . . I was so

---

[13] *Collected Papers.* See also Fölsing.

happy to have my lover for myself again for a while, the more so as I saw that he was just as happy.

Soon after this trip Mileva must have told Einstein that she was pregnant, since toward the end of May 1901 his letters to her start referring obliquely to the happy event, assuring her that he would stand by her and all would be well. Typically, though, those letters are much more excited about a wonderful new piece of physics Einstein has heard of. The German physicist Philipp Lenard had discovered that electrons could be knocked out of the surface of a metal by shining ultraviolet light onto it. Strangely, he had also discovered that the energy of the electrons ejected from the metal surface did not depend on how bright the light was. Whether it was faint or dim the electrons always came out with the same energy (essentially, the same speed). How could this be? Just four years later, Einstein would explain the phenomenon; and in 1922, he would receive the Nobel Prize for his explanation, with profound effects on the lives of himself and Mileva. But in 1901, the promise of the spring soon faded once again into uncertainty about their futures.

Einstein quite liked his brief time as a teacher, and was especially pleased to discover that after five or six hours spent at the task (schools started early in Switzerland) he was still fresh enough to work in the library or on interesting problems at home, so that he could continue his scientific work even if he didn't have any official connection with a university. And on Sundays he could take the train into Zurich to see Mileva. But

when the temporary job ended, and the reality of the implications of Mileva's pregnancy began to sink in, the future once
again looked less promising, with no sign yet of the promised vacancy at the Bern patent office. In July, Mileva failed her exams
again, almost as badly as she had done the year before. Abandon-
ing all hope of a PhD, pregnant, and depressed, she went
home alone to see her parents and break the unwelcome news.
Einstein took on an unsatisfactory job as a private tutor just
to make ends meet, and devoted all the time he could to writing a new dissertation, developing his ideas concerning the
forces between molecules. He hoped to use this to obtain a
PhD, which he now regarded as his best chance of getting an
academic job.

In the autumn, Mileva retuned to Switzerland, but couldn't
stay with Einstein, or be seen with him, since her now visible
pregnancy would have compromised his position as a respectable teacher. She stayed in a nearby town, while he made a
succession of excuses that he could not find time to get away
and see her. They had little contact, and once again she went
back to her parents. In November, Einstein completed his dissertation and submitted it to the University of Zurich, scraping
up the fee of 230 francs required for it to be considered. Professor Alfred Kleiner read the dissertation, but infuriated Einstein
first by taking until well into the new year to comment on it,
then advising the young man to withdraw it before it went any
further. A furious Einstein did so (which at least meant he got
his 230 francs back), and ranted to his friends about the incom-
petence of all professors, especially Kleiner. But the professor

probably did Einstein a favor. No copy of the dissertation survives, and we cannot know what was wrong with it, but judging from the papers in *Annalen der Physik* on which it is supposed to have been based, Kleiner probably had a point. Einstein himself later described the paper as "worthless."[14]

In any case, Einstein's anger was a little assuaged by news he received while Kleiner was still considering the dissertation. On December 11 he heard from Grossmann that the vacancy in the patent office was about to be advertised. Uniquely in the history of the patent office up to that time, the ad specified that applicants should have a university education with a "specifically physical direction"; before 1902, there were no physicists in the Swiss patent office. Impulsively, as well as applying for the post, Einstein gave up his job and moved to Bern in January 1902, blissfully unconcerned that even if he got the job he would not be starting immediately. At about the same time, he became a father; after a difficult (indeed, life-threatening) labor Mileva gave birth to a girl, Lieserl.

Although their correspondence clearly shows that the couple initially intended to keep their daughter, there was no prospect of Mileva bringing her to Switzerland as soon as her health permitted. A respectable Swiss civil servant could not possibly be seen to be flouting conventional morality—and conventional morality in Bern at the time was so strict that women were legally forbidden to smoke in the street. There was no prospect of marriage until he had the patent office job—and there would be no

---

[14] See Fölsing.

prospect of the job if the authorities knew about his relationship with Mileva. So, in the short term at least, Einstein would be living as an impoverished but independent bachelor in Bern. It was almost an echo of his early days as a student in Zurich.

The Swiss bureaucracy only moved slowly toward the process of carrying out interviews and appointing new patent officers to fill the two vacancies that were becoming available, and meanwhile Einstein had to live. He decided to make a little money by private teaching, helping students at the University of Bern, and advertised his services in the local paper, picking up a couple of students willing to pay the modest fee he requested. One of these men, Louis Chavan, soon became a good friend. Einstein already had some contacts in Bern. An old schoolmate from Aarau, Hans Frösch, was now studying medicine at the university, and Paul Winteler, one of Marie's brothers, was studying law. Another old friend, Max Talmey, was traveling in northern Italy in the spring of 1902; after visiting Einstein's parents in Milan he called in to see Einstein himself in Bern. Talmey found Einstein living in a "small, poorly furnished room"[15] and struggling to make ends meet. But if things were so hard financially that Einstein wasn't even getting enough to eat, socially and intellectually he was having the time of his life.

During the Easter vacation, Einstein ran the ad offering his services as a tutor again. This time, one of the responses came from Maurice Solovine, a twenty-six-year-old Romanian student from a wealthy family who was fascinated by the big ideas in

---

[15] Article in the *New York Post,* February 25, 1931.

physics and philosophy, but had no clear idea of what direction to follow. In him, Einstein found a kindred spirit, and the idea of Solovine paying for his education was soon forgotten as they became firm friends and met frequently to discuss the big issues of the day. Early in the summer, they were joined by Conrad Habicht, a mathematics student, and took to calling themselves "the Olympia Academy," meeting regularly in the evenings and working their way steadily through books on the big ideas in philosophy and physics. Among the most influential of these works on Einstein himself were those by David Hume, Ernst Mach, and Henri Poincaré. It was Poincaré who introduced Einstein to the concept of non-Euclidean geometries, mathematically self-consistent and logical versions of geometry in which, for example, the angles of a triangle do not add up to 180° and parallel lines can either meet or diverge from one another.[16] Poincaré's book *Science and Hypothesis,* which the "Academy" studied and discussed in detail, poses an interesting question about the relationship between these geometries and the world we live in. Poincaré asked his readers what would happen if astronomers discovered that a pair of parallel light rays traveling through space eventually converged on one another. Would they conclude that space obeyed non-Euclidian geometry? Or would they conclude that some unknown force was bending the light rays? Poincaré had no doubt that the answer would be in favor of the unknown force.

---

[16] Coincidentally, Marcel Grossmann was studying non-Euclidean geometry for his PhD at around the same time.

While Albert enjoyed himself in poverty in Bern, Mileva had returned to Zurich, without her baby, who was left with her relatives or a friend back home. Einstein's visits were rare, even when she moved to a small town closer to Bern, and in spite of the warmth of his letters to her it seems likely that the relationship might have ended in the way his relationship with Marie Winteler had ended, if it had not been for the child and the sense of duty that he felt toward the woman that he had, in what was still the language of the day, "ruined."

Mileva's prospects brightened in May 1902, when Einstein was at last called for an interview at the patent office and duly offered the post of "Technical Expert III Class." The other person appointed at the same time was an engineer, Heinrich Schenk. The patent office had recognized the need for a physicist because so much of their work now concerned inventions based on the application of electromagnetism; but Einstein's lack of experience in technical matters meant that he was initially appointed on probation, at a salary of 3,500 francs a year (roughly twice what he would have got in a university assistantship). Although he took up the post in June 1902, it wasn't until September 1904 that his appointment was confirmed as permanent and his salary increased to 3,900 francs. But even the initial salary was a fortune to Einstein, who happily wrote to Mileva that now he could abandon the "annoying business of starving."[17]

There was now no financial impediment to the marriage.

---

[17] *Collected Papers.*

But there was still strong opposition from Einstein's parents, who knew about the baby. Einstein's mother, in particular, regarded Mileva as a scheming foreign hussy who was trying to trap her son; the situation wasn't helped by the fact that even though the Einsteins were secular Jews, nobody in the family had married a non-Jew before. In spite of his independent nature and rebellious streak, Einstein couldn't bring himself to defy them on this occasion. His reluctance may have been partly due to a desire not to cause them more grief at a time when the family was already under stress. The latest business hadn't actually gone bust, but it was clear that it would never be profitable enough to repay the huge debts that the Einsteins owed to their relatives, who had begun pressing for the money. In addition, Einstein's father was ill with heart trouble, undoubtedly exacerbated because of the stress caused by his years as an unsuccessful businessman. In October 1902, his heart finally gave up the struggle; Hermann was just fifty-five.

According to Einstein's sister,[18] when Hermann was on his deathbed, he relented and gave Einstein formal permission to marry Mileva. Ironically, though, his death left Einstein rather less able to support a wife, since he began sending part of his salary to his mother each month, to help her to pay off the debts her husband had left her with. Nevertheless, toward the end of the year Mileva moved to Bern, and on January 6, 1903, the couple were married at the register office. The only witnesses were the other two "Olympians," Habicht and Solovine.

---

[18] See Overbye.

Much later, Einstein would admit to his biographer Carl Seelig that he had married from a "sense of duty" in spite of his feelings of "internal resistance." For her part, Mileva still seems to have regarded it as a love match, and to have happily taken on the domestic role, cooking and looking after her new husband, while tolerating his friends. The only problem, from her point of view, was what to do about Lieserl.

The couple settled into a comfortable existence in their apartment in Bern. Albert wrote to his old friend Besso that "[I] am living a very pleasant, cosy life with my wife. She takes excellent care of everything, cooks well, and is always cheerful."[19] While Mileva wrote to her friend Helene Savic,[20] "[Albert] is my only companion and society and I am happiest when he is beside me."

But she was not his "only companion and society." He still had his work at the patent office, which he enjoyed, and his friends, including the Olympians. And he had his research, which was now developing along extremely promising lines, which he carried out in his own time and also when things were quiet in the patent office. Between 1902 and 1904 Einstein produced a series of three papers, all published in the *Annalen der Physik,* which came close to making him a name as an important member of the scientific community. Working completely on his own, he found how to interpret the laws of thermodynamics mathematically entirely in terms of the statistical behavior of a myriad of tiny particles, laying the foundations of

---

[19] *Collected Papers.*
[20] See Overbye.

statistical mechanics. The only snag was, entirely unknown to Einstein, the American Josiah Willard Gibbs had beaten him to it and published what became the standard work on statistical mechanics in 1902. Their two approaches were almost the same, but Einstein only came across Gibbs's book in 1905, after it had been translated into German. The fact that two people independently came up with the foundations of statistical mechanics at about the same time is not surprising, since this step was (for anyone clever enough to take it) a logical progression from the work of the pioneers such as Maxwell and Boltzmann. It was just Einstein's bad luck that on this occasion he came second; but anyone who read and understood those papers in the *Annalen der Physik* can have been left in little doubt that the author was a scientist of note.

We don't know what discussions took place between Albert and Mileva in the first months of their marriage concerning Lieserl, but in August Mileva set off back to her homeland to sort the situation out. She was away for a month, during which she found out that she was pregnant again, and returned without the little girl. Historians have scoured the fragmentary correspondence that survives from the period, and the official records, in an effort to find out what happened to Lieserl, but in effect she just disappeared from sight in September 1903. The consensus is that she was adopted, most probably given up formally to strangers or possibly informally taken into the family of Mileva's friend Helene Savic and given another name. A reference to scarlet fever in one of Albert's letters to Mileva also raises the possibility that she died in infancy, as so many children did at the time. Whatever Lieserl's fate, when Mileva returned to Bern in the autumn of 1903 it was essentially a fresh start, crowned by the arrival of Einstein's first son, Hans Albert, on May 14, 1904.

By then, there were major changes in Einstein's circle of friends. Habicht and Solovine had finished their studies and moved on, but by that time Einstein had a more than adequate scientific sounding board to replace them. His friend Michele Besso had been finding it hard to make a living as a freelance engineer, and when a vacancy for technical expert (II class) came up at the patent office, Einstein drew Besso's attention to it. Besso applied and got the job, starting in the summer of 1904. He was a grade above Einstein, and earned an annual salary of 4,800 francs (Einstein's raise to 3,900 francs would not come through until September), but this was entirely appropriate to his age and experience.

Now Einstein had a companion he could talk to about science during breaks at work, while walking to and from the patent office, and during his so-called leisure time. But there was very little real leisure time. Contemporary accounts describe how even when Einstein was pushing Hans Albert about in his baby carriage on a Sunday afternoon stroll, he would have his pipe in his mouth and a notepad resting on the carriage, ready to write down his thoughts when inspiration struck. Thanks to Mileva, he had no domestic worries at all, and never had to concern himself with routine details like cooking and cleaning. It was in this sense that she would make a major contribution to his miraculous year, and although she might have dreamed of sharing a scientific partnership with Albert like that of Marie and Pierre Curie (who won the Nobel Prize together in 1903), it was not to be.

With the breakup of the Olympians and the arrival of his son and Besso on the scene, and with three solid scientific papers under his belt, Einstein now seems to have taken himself more seriously. Before, ideas had been tossed about in discussions over coffee and wine, but few of them had been followed through rigorously. Now, he buckled down to work through some of the ideas that had been floating around in his head for years. First, he would complete a new dissertation, making use of his growing understanding of statistical processes, and make a proper try for that elusive PhD. And then, there were three other ideas he had, concerning the reality of atoms, Planck's light quanta, and the old puzzle about what the world would look like if you could travel at the speed of light. By the end of 1904 the scene was set, although even Einstein cannot have sensed it, for the greatest outpouring of scientific creativity since the time of Isaac Newton.

PAUL HABICHT, MAURICE SOLOUINE, AND ALBERT
EINSTEIN AT BERN. THESE THREE CLOSE FRIENDS
NAMED THEMSELVES "THE OLYMPIAN ACADEMY."

CHAPTER TWO

The Annus Mirabilis

In March 1905, Albert Einstein celebrated his twenty-sixth birthday; two months later, his son, Hans Albert, celebrated his first birthday. That year, Einstein was living in a settled household with a steady job, his close friend Besso to talk to about scientific matters, and a routine for working on his own scientific projects in his own time. This didn't leave much time for his family, and although he was a loving father he would hardly be regarded as a role model today. The problem of Lieserl had been resolved, for better or for worse, and once it was resolved his daughter seems to have been dismissed from his mind. Mileva, for all her onetime scientific aspirations, was essentially a conventional little housewife, looking after her husband so that he could concentrate on higher things. And as for being a father, rather than suffering the distractions most new fathers go through, Einstein just ignored them, the way he would ignore anything that threatened to disturb his scientific work throughout his life.[1] Several accounts recall how on those occasions when he was supposed to be looking after the baby he might be found with his pipe in his mouth, rocking the cradle with one hand, while writing out calculations on his ubiquitous notepad with the other. This ability to switch off from the distractions of the world around him and to concentrate on the problems he was interested in was a major reason why he was able to produce such an outpouring of papers in 1905, and to maintain a high standard of scientific achievement for many more years to come.

The first hint of what Einstein had been up to in the early months of 1905 came in a letter he wrote at the end of May to his friend Conrad Habicht:[2]

I promise you four papers, the first ... deals with radiation and the energetic properties of light and is very revolutionary, as you will see ... The second paper is a determination of the true size of atoms by way of the diffusion and internal friction of diluted liquid solutions of neutral substances. The third proves that, on the assumption of the molecular theory of heat, particles of the order

---

[1] The story that in later life he gave up wearing socks so as not to have the bother of finding clean ones to wear is true, and typical.
[2] See Seelig.

of magnitude of $\frac{1}{1000}$ millimeters suspended in liquids must already perform an observable disordered movement, caused by thermal motion. Movements of small inanimate suspended bodies have in fact been observed by the physiologists and called by them "Brownian molecular movement." The fourth paper is at the draft stage and is an electrodynamics of moving bodies, applying a modification of the theory of space and time; the purely kinematic part of this paper is certain to interest you.

That has to be one of the most remarkable letters in the history of science. The first paper Einstein mentions established the reality of light quanta (what we now call photons), and was so revolutionary that it earned him the Nobel Prize—although because it was so revolutionary it took sixteen years for the rest of the scientific community to catch up with him and make that award. The second paper formed the basis of his doctoral dissertation. The third proved the reality of atoms. And the fourth presented a bemused world with the special theory of relativity.

We will describe these pieces of work in a slightly different order, starting with the doctoral thesis. One good reason for doing this is that this was the least revolutionary of the four papers, and Einstein knew it. When he decided to make a second attempt at obtaining a PhD from the University of Zurich, Einstein didn't set out on a specific new project to achieve that goal, but simply seems to have looked at the various projects he had in hand and chosen the most straightforward one, based on solid, traditional methods, that wouldn't tax the imagination of the professors at the university too much. He was also careful to choose a piece of work based on experimental observations, although he didn't carry out the experiments himself. There is a story, originating with Einstein's sister, Maja,[3] that Einstein first offered his paper on the electrodynamics of moving bodies (the special theory of relativity) and it was rejected because the examiners didn't understand it; but this seems to be a myth. There is no evidence of such a rejected application, Einstein had more sense than to baffle the examiners in this way, and in any case the paper was the last of the four to be completed, as the letter to Habicht shows.

Einstein actually completed the paper that became his dissertation, prosaically titled *A New Determination of Molecular Dimensions,* at the end of April; but he didn't submit it to the University of Zurich until July 20, 1905. The delay may have been because he was busy working on his other ideas in the spring, and didn't make a final decision about which paper to submit until then. Although the title of the dissertation focuses on the sizes of molecules, the technique Einstein describes actually also gives a measurement of the number of molecules (or atoms) present, in this case in a solution. This is typical of the kinds of methods used to estimate the numbers and sizes of atoms and molecules in the nineteenth and early twentieth centuries, and shows Einstein building on what has gone before, rather than leaping off in a new direction.

---

[3] See, for example, Fölsing.

The problem was that in mathematical descriptions of the behavior of atoms and molecules, both the sizes and the numbers of these particles appear in the equations. As we all learned in school, if you have one equation involving one unknown quantity (usually denoted by x), you can solve the equation to find a value for the unknown quantity. But if you have just one equation involving two unknowns (x and y), you cannot find out what the values of the unknowns are. To do that, you need two different equations each involving x and y. So all the "classical" methods for determining the sizes of molecules *and* the numbers of molecules in a certain amount of matter depended on using two equations to work out the two unknowns.

The number that comes into these calculations was called Avogadro's number,[4] after the Italian who came up with the idea in 1811. It is the number of particles (atoms or molecules) contained in an amount of material whose weight in grams is numerically equal to the atomic (or molecular) weight of the substance. The atomic weight of carbon, for example, is 12, so 12 grams of carbon contains Avogadro's number of atoms. The atomic weight of hydrogen is 1, but each hydrogen molecule contains two atoms, so its molecular weight is 2. So 2 grams of hydrogen gas also contains Avogadro's number of molecules—and so on.

One early attempt to work out the value of this number, and simultaneously the sizes of molecules, was made by the German

---

[4] This number is called Avogadro's "constant" today, but we shall stick with the name familiar in Einstein's day.

Johann Loschmidt in the mid-1860s. His calculations involved the average distance traveled by particles in a gas between collisions with one another (called the "mean free path") and the fraction of the volume of the gas actually occupied by the molecules themselves. He reasoned that in a liquid all the molecules must be touching each other with no gaps in between, so measuring the density of the liquid, which depends on the number and size of the molecules present, would tell you the volume occupied by Avogadro's number of molecules. When the liquid is heated to become a gas, the actual molecules must still occupy the same volume as the original liquid, but now with lots of empty space between them. It's only in the gas that the mean free path comes into the calculations.

Loschmidt carried out his calculations for air, which is almost completely a mixture of oxygen and nitrogen, and had to use estimates for the density of liquid nitrogen and liquid oxygen that were not as accurate as modern measurements. He combined these with calculations of the mean free path, which also depends on the number and size of the molecules present, based on measurements of the way the pressure exerted by air changes when it is squeezed into a smaller volume. He found that a typical molecule of air must be a few millionths of a millimeter across, and estimated Avogadro's number to be 0.5×10^{23}—which means a 5 followed by 22 zeroes, or fifty thousand billion billion.

Einstein's approach to the problem was in the same spirit of solving two different equations simultaneously, but used a very different kind of physical system. He realized that the sizes of molecules (and Avogadro's number) could be inferred from measurements that had already been carried out on the behavior of solutions of sugar in water. But, as we have said, he didn't do the experiments. What was new about Einstein's work, and justified the award of his PhD, was the mathematical way he calculated how molecules of sugar would behave in such solution, and how this would affect the measurable properties of the solution. What was particularly clever about the work wasn't that it gave a value for Avogadro's number or the size of molecules—the techniques based on the kinetic theory of gases, such as Loschmidt's method, had already done that. Einstein's special contribution was to find a way to get results as good as those obtained from the kinetic theory of gases using liquids alone. Previously, estimates based on studies of liquids had been very rough and ready. Along the way, as we shall see, he developed techniques with widespread applications everywhere that industry uses suspensions of particles in liquids.

The technique depended on the fact that sugar molecules are much larger than molecules of water. In fact, as Einstein realized, because some water molecules actually attach themselves to the sugar molecules in the solution, the effective size of the sugar molecules is even bigger, which makes the assumptions used in his calculations even more accurate.

It is easy to describe the thinking behind those calculations. When something is dissolved in water, the viscosity of the solution—its stickiness—increases. By assuming that each sugar molecule is a large sphere embedded in a sea of much smaller water molecules, Einstein was able to work out an equation that related this change in viscosity (which can be measured) to the total volume of the fluid occupied by the sugar (which depends on two unknown quantities, the size of each sugar molecule and the number of sugar molecules present). Because experimenters always measure the weight of sugar (or other stuff) being added to the solution, the number of molecules present can always be expressed in terms of Avogadro's number.

Then Einstein looked at the way sugar diffuses through water, and calculated the force acting on a single sugar molecule as it moves through the sea of water molecules. This could be related to another measurable property of the solution, called its osmotic pressure, through another equation which itself depended on both the number of sugar molecules present and their size. So Einstein derived two equations, each of which included the two unknown quantities, Avogadro's number and the size of a sugar molecule, and each of which was directly related to measurable properties of the solution, its viscosity and the osmotic pressure.

Once he had that pair of equations, it was a simple matter to plug in the numbers for viscosity and osmotic pressure that were already well known and had been published in standard tables listing the properties of such solutions. For the record, the "answers" that came out of the equations were that sugar molecules (which are much bigger than molecules of nitrogen or oxygen) have a radius of about 9.9×10^{-8} cm (9.9 hundred-millionths of a centimeter) and Avogadro's number is 2.1×10^{23} (210 thousand billion billion). This pretty much agrees with Loschmidt's estimate;[5] but the impressive part of the dissertation is the bit we can't put into words, the sophisticated mathematical techniques that Einstein used to deduce the relevant equations.

The best way to appreciate just how good the math was is to look at what the professors who examined the work said in their official report to the University of Zurich. Alfred Kleiner commented that "the arguments and calculations to be carried out are among the most difficult in hydrodynamics and could be approached only by someone who possesses understanding and talent for the treatment of mathematical and physical problems. . . . Herr Einstein has provided evidence that he is capable of occupying himself successfully with scientific problems," while his colleague Heinrich Burkhardt said that "the mode of treatment demonstrates fundamental mastery of the relevant mathematical methods."[6] Einstein himself later told his biographer Carl Seelig that the only official comment he had received on the dissertation was that it was too short, and that in response he had added a single sentence, whereupon it was accepted. Even though the dissertation itself was officially accepted by the University of Zurich in early August 1905, it took until January 15, 1906, to complete all the various formalities required by the university for the degree to be conferred. So all the papers

---

[5] Remember that what matters is not so much the number on the front, 0.5 or 2.1, as the agreement of the number of "powers of ten" in the exponent, 23.
[6] *Collected Papers*; Kleiner translation from Fölsing, Burkhardt translation from Stachel.

he completed during the annus mirabilis were the work of simple Herr Einstein, not Herr Doktor Einstein.

Just after the thesis was accepted, Einstein submitted a slightly revised version to the *Annalen der Physik*, but publication was delayed until 1906 because the editor of the journal, Paul Drude, knew of some more accurate and up-to-date measurements of the properties of sugar solutions than the ones Einstein had used. When he asked Einstein to take account of these data, the result was a slight change in the numbers he came up with—in the right direction, we now know. Even that wasn't the end of the story, because Einstein's paper eventually encouraged other experimenters to measure the relevant properties of these and other solutions even more accurately. It also turned out that Einstein had made a minor error in his calculations, which had not been spotted by either of his examiners or by Drude. The final, definitive version of Einstein's method for calculating Avogadro's number from the properties of sugar solutions only appeared in the *Annalen der Physik* in 1911, and gave a value of 6.56×10^{23}, which is very close to the accepted modern value, 6.02×10^{23}.

The saga of Einstein's doctoral dissertation did not end there, however. In scientific terms, this is the most mundane of the papers he wrote during the annus mirabilis. But it has one curious distinction. It became far more widely quoted than any of his truly revolutionary papers from the same year.

One way in which scientists measure the value of scientific papers is to record how often they are referred to in other scientific papers. This is by no means a perfect system, as witnessed by the fact that Einstein's original paper on the special theory of relativity has been very seldom referred to. The reason for that, of course, is the content of the paper quickly became part of the established fabric of science, something taught from textbooks that "everybody knows," so that few scientists have even read the paper, let alone cited it.

By contrast, the paper based on the doctoral dissertation has been very widely cited. Just how widely was brought home in 1979, as part of the celebrations to mark the centenary of Einstein's birth. Two researchers[7] carried out a survey of the citations received not just by Einstein's papers but by all the papers in science (what they called the "exact" sciences, like physics and chemistry) published before 1912. The twist was that they only looked at citations in papers that had themselves been published between 1961 and 1975, so they came up with a list of all the papers that were still important enough to be quoted at least fifty years after they had originally been published. Out of the top eleven "most cited" papers in this survey, four were by Einstein (no other scientist had more than one paper in the top eleven). And top of the four papers by Einstein came his doctoral dissertation.

Why was this seemingly mundane paper cited so often between 1961 and 1975? Simply because it is mundane. It deals with practical things important in the everyday world—the behavior of fluids with particles suspended in them. The equations

---

[7] Tony Cawkell and Eugene Garfield, contribution to *Einstein: The First Hundred Years,* ed. M. Goldsmith, A. Mackay & J. Woudhuysen, Pergamon, Oxford, 1980.

Einstein derived are relevant in, among other places, the dairy industry, where it is important to understand and predict the behavior of milk during the process of making cheese; to the study of pollution and the way tiny particles called aerosols get spread through the atmosphere; and to problems involving the behavior of cement being transported in liquid form, and the design of the trucks to carry the cement. The work Einstein did for his doctorate in 1905 turned out to be of widespread importance in many practical applications in the second half of the twentieth century, and is still relevant today, a hundred years after the dissertation was written.

For the same reasons, the second most cited of Einstein's papers in that 1979 survey did not concern the special theory of relativity or quantum physics (indeed, neither of those papers even made the top four), but the phenomenon known as Brownian motion. Appropriately, it was the paper on Brownian motion that Einstein returned to as soon as he had finished drafting what would become his doctoral dissertation, at the end of April 1905. On May 11, Paul Drude received a paper with the splendid title *On the Motion of Small Particles Suspended in Liquids at Rest Required by the Molecular-Kinetic Theory of Heat*.[8] He had no hesitation in accepting it for publication.

Brownian motion got its name from the Scottish botanist Robert Brown, who first studied the phenomenon in detail

----

[8] You sometimes see slightly different versions of the titles of Einstein's papers. Ours, and all our quotes from Einstein's papers, are taken from Stachel, which is the most accessible source.

in 1827. Intriguingly, though, Einstein wasn't trying to *explain* Brownian motion in this paper; indeed, in a sense he was *predicting* it, on the basis of his statistical approach to the kinetic theory, honed in the series of three papers we mentioned earlier. That's why the term "Brownian motion" doesn't appear in the title of the paper. In the first paragraph of the paper, Einstein says:

> It is possible that the motions to be discussed here are identical with so-called Brownian molecular motion; however, the data available to me on the latter are so imprecise that I could not form a judgment on the question.

But since we now know enough to form that judgment, it makes sense to introduce this aspect of Einstein's work by looking at just what it was that Robert Brown discovered.

Even before Brown's time, people had noticed the way tiny grains of material, notably pollen, seem to dance about in a jittery kind of motion, something like running on the spot, when they are suspended in a liquid such as water and observed through the microscope. So Brown didn't discover Brownian motion. Before Brown's work, however, the obvious explanation for this motion seemed to be that the particles were alive— after all, pollen grains are a kind of plant equivalent to the sperm cells in animals, and if sperm can move under their own steam, why shouldn't pollen? When Brown began his detailed studies in the summer of 1827 (the results were published in 1828), he thought that this was the most probable explanation. But then he made the next logical step. He took a series of clearly inanimate materials, such as ground-up fragments of glass and granite, and suspended them in water. He found exactly the same behavior for these definitely nonliving materials, proving that the motion of a particle in suspension has nothing to do with any mysterious life force. "These motions," he wrote in that 1828 paper,[9] "were such as to satisfy me, after frequently repeated observation, that they arose neither from current in the fluid nor from its gradual evaporation, but belonged to the particle itself." It was this discovery, a result of the truly scientific way he went about his work, that meant his name would be forever associated with the phenomenon.

But if a life force wasn't causing the motion, what was? Over the next few decades, people considered the possibility that convection currents might be involved (in spite of Brown's comments), or electrical effects, or the same force that caused capillary action, and other more or less wild ideas. The key experimental discoveries were that the speed of this jiggling increased if the temperature of the water increased, and was less for bigger particles. Combining this with the ideas of the kinetic theory gave rise to the suggestion that the particles were being bombarded by the molecules in the water, and were being jerked about in response to the kicks they received from individual molecules. But in order for a single molecule to produce a visible shift in a pollen grain or a speck of granite dust, the molecule would either have to be impossibly big or traveling impossibly fast.

This was more or less where the puzzle of Brownian motion

---

[9] In the *Philosophical Magazine*.

stood at the beginning of the twentieth century. It is clear from his writings, though, that Einstein had not read up on all of these developments, and was not up to date on the subject. He was aware of the phenomenon of Brownian motion, but his theoretical studies of how particles suspended in liquids ought to move were not specifically intended to explain that phenomenon. Rather, they were a logical development from the work in his doctoral dissertation.

As Einstein has told us, what he was really interested in at that time was proving the reality of atoms and molecules. He was completely convinced of the validity of the kinetic theory of heat, and saw in this extension of his PhD work a way to convince others as well. For this purpose, the distinction between atoms and molecules is of no significance. Atoms are the fundamental component of elements, such as hydrogen and oxygen, and molecules are the basic components of compound substances, such as water (where two hydrogen atoms combine with one oxygen atom in each molecule of water).

At its simplest, the kinetic theory says that everything is made of tiny particles (atoms or molecules), which can be regarded as little, hard spheres. In a solid, the little spheres are packed closely together and do not move past one another. In a liquid, the little spheres buffet each other and slide past one another like people moving through a dense crowd, but they are still essentially touching all their neighbors. In a gas, the little spheres fly freely through empty space, bouncing off each other and the walls of any container they are in. The hotter a substance is, the faster the spheres move, which explains the transition from solid to liquid to gas as a substance is heated, and from gas back to liquid and then solid when it cools.

In the paper that became his doctoral dissertation, Einstein had already used the idea that molecules of sugar dissolved in water are being bombarded by water molecules from all sides, and that the way the sugar molecules move through the sea of water molecules affects measurable properties of the solution, its viscosity, and its osmotic pressure. The success of Einstein's results from that paper already provided powerful circumstantial evidence in favor of the kinetic theory; but even that was not direct proof that atoms and molecules exist. To obtain that, the effects of the bombardment by water molecules had to be scaled up somehow, to become visible, at least under the microscope. A pollen grain, tiny though it is by any human standard (about a thousandth of a millimeter across), is enormously much bigger than a water molecule (measured in millionths of a millimeter), or even a molecule of sugar. But Einstein made the huge mental leap of realizing that as far as the behavior of particles suspended in liquids was concerned, this was the *only* difference that mattered between a pollen grain (or a fragment of granite) and a sugar molecule. In what we shall refer to as the Brownian motion paper, he said:

According to [the kinetic theory], a dissolved molecule differs from a suspended body *only* in size, and it is difficult to see why suspended bodies should not produce the same osmotic pressure as an equal number of dissolved molecules. We have to assume that the suspended bodies perform an irregular, albeit very slow, motion in the liquid due to the liquid's molecular motion.

And he went on to calculate both that osmotic pressure and the nature of that irregular motion.

The osmotic pressure that comes into both this work and the doctoral dissertation is a curious phenomenon worth describing in a little more detail. If you have a container of water (or some other liquid), like a fish tank, it can be divided into two by putting in a barrier that has tiny holes in it, just big enough for water molecules to pass through. If you do this, the water can get from one side (either side) of the barrier to the other. This is known as osmosis. But if you now dissolve something (such as sugar) in the water on one side of the tank, the dissolved molecules are too big to get through the holes. The barrier in such a setup is then called a semipermeable membrane, because it lets some molecules through but not others.[10] This is where things get interesting.

You now have a solution of sugar in water on one side of the membrane, and pure water on the other side. The result is a pressure that moves water molecules from one side of the membrane to the other. You might guess (most people do, the first time they come across this) that the presence of the sugar molecules pushes water out of that half of the tank, making the

---

[10] All of this works for other solutions as well, of course, but we shall stick with sugar and water because that was the example Einstein used in his dissertation.

solution stronger and raising the level of the water on the other side of the barrier. In fact, just the opposite happens. Water from the pure side of the tank passes through the membrane, making the sugar solution more dilute, and increasing the height of the liquid on the side of the barrier where the sugar is. The process only stops when the pressure of the extra height of liquid (the osmotic pressure) is enough to stop the flow of water molecules through the membrane.

This counterintuitive behavior is an example of the famous second law of thermodynamics at work. We don't have space to go into all the details here, but the relevant point, at the heart of that law, is that natural processes tend to even out irregularities in the universe. On a grand scale, the Sun and stars are pouring out heat into the cold universe; on a more homely scale, an ice cube in a glass of water melts, evening things out to produce an amorphous liquid. In the example of osmotic pressure, the water molecules that move into the sugar solution make the solution more dilute, more like the pure water, so that there is less contrast between the fluids on opposite sides of the semipermeable membrane.

In the Brownian motion paper, Einstein first covered some ground similar to parts of his doctoral dissertation, but using a different (and rather more elegant) mathematical approach. His calculations involved the relationship between osmotic pressure, viscosity, and the way individual particles suspended in the liquid diffuse through the sea of molecules surrounding them. But this time, he was describing the behavior of particles big enough to see under the microscope.

The way Einstein set about his work was, though, just as important as the results he obtained. He realized that the kick produced by a single molecule hitting a particle as large as a pollen grain could not produce a measurable shift in the position of the large particle. But the large particle is constantly being bombarded by molecules, from all sides. On average, the kicks from one side are balanced by the kicks from the opposite side, so you might not expect the large particle to move at all. Einstein realized, however, that the important words are "on average." If you take a very small time interval, then just by chance at that instant the particle will be receiving more kicks on one side and fewer on another. The combined effect will be to shift the particle by a minute amount in the direction of least resistance. Then, in the next instant the pattern will change, and the particle will shift in another direction, and so on. Einstein's special insight into the nature of this kind of statistical fluctuation was that what happens during each of these small time intervals is entirely independent of what happens in any other time interval, even the one just before the one being considered.

Because of this independence, and the statistical nature of the fluctuations, the particle doesn't simply move to and fro around the same spot that it started from. Nor does it keep moving in one direction. Einstein discovered that it gradually moves farther and farther away from its starting point, but following a zigzag path that has become known as a random walk. He showed that wherever the particle starts from, the distance it moves away from its starting point depends on the square root of the time that has passed. So if it moves a certain distance in one second, it will move twice as far in four seconds (because 2 is the square root of 4), four times as far in 16 seconds, and so on. But it doesn't keep going in the same direction—after four seconds it will be twice as far away from the start as it was after one second, but in a random and unpredictable direction.

This is called a "root mean square" displacement, and the equation Einstein worked out for the displacement involves the temperature of the liquid, its viscosity, the radius of the particle and Avogadro's number. He used this equation and a value for Avogadro's number inferred from other experiments to predict that a particle with a diameter if 0.001 mm in water at a temperature of 17°C would shift a distance of 6 millionths of a meter from its starting point in one minute. But he also realized that if the predicted displacement could be measured accurately enough, the same equation could be used the other way around, to give a value for Avogadro's number.

The prediction provided a classic example of the scientific method at work, since measuring the way a particle moved away from its starting point would answer the question of whether the theory it was based on was right or wrong. As Einstein put it in his paper:

> If the prediction of this motion were to be proved wrong, this fact would provide a far-reaching argument against the molecular-kinetic conception of heat.

It wasn't proved wrong. Although Einstein didn't know it when he wrote the paper, in the early 1900s microscopists were already developing improved instruments, known as ultramicroscopes, that would be able to measure the kind of motion he was describing accurately enough to test the prediction.

The Brownian motion paper was published in July 1905, and almost immediately Henry Siedentopf, a German working with the new ultramicroscope, wrote to Einstein to tell him that the kind of motion described in his paper almost certainly was Brownian motion. It still wasn't possible at that time to test Einstein's detailed predictions, but he was sufficiently encouraged to write another paper, this time plainly titled *On the Theory of the Brownian Motion*, which he sent off to the *Annalen der Physik* in December; it was published in 1906. In this paper he developed his ideas further and also predicted that particles suspended in a liquid would experience a rotary movement, dubbed Brownian rotation, although he did not expect this to be observable.

It was extremely difficult to make the observations required with enough accuracy to test Einstein's predictions, and several researchers tried and failed over the next couple of years; but in 1908 the French physicist Jean-Baptiste Perrin finally succeeded. Instead of trying to measure the displacement of individual tiny particles from their starting point, Perrin used another result that had by then emerged from Einstein's theoretical model, which predicted the way particles suspended in a solution would be arranged vertically.

In such a suspension, the particles would be tugged downward by gravity, gradually sinking to the bottom of the liquid. But superimposed on this very slow downward drift would be Einstein's random walk. The overall effect would be a vertical distribution of particles, with more at the bottom and fewer at the top, obeying a precise mathematical law (a specific exponential, decreasing with increasing height). Perrin's results exactly matched the predictions from Einstein's theory, and he even went one better by measuring the Brownian rotation that Einstein had predicted before anyone had seen it. He also used the observations to make an accurate measurement of the value of Avogadro's number.

The whole package finally established the reality of atoms and molecules, and the validity if the kinetic theory, silencing the few (by then, very few) remaining doubters. This work was so important that Perrin received the Nobel Prize "in particular for his discovery of the equilibrium of sedimentation," as the citation put it, in 1926.

But Einstein's work had even wider applications. The kind of statistical methods he used, coupled with the idea of random events occurring independently in individual tiny intervals of time, proved fruitful across a whole range of topics in physics. In his second paper on Brownian motion (still written in, though not published in, the annus mirabilis), Einstein had pointed out the possibility of applying the same approach to the study of fluctuations in electric circuits (the phenomenon now known as "noise"), and in years to come the technique would be widely applied in the new field of quantum physics. There, for example, exactly the same combination of statistical effects and random changes occurring in independent time intervals leads to an understanding of the nature of the half-life associated with radioactive processes.

This link with quantum physics is particularly appropriate, because the next paper we shall discuss from Einstein's miraculous year saw him laying one of the foundation stones on which the whole edifice of quantum theory was built. Curiously, though, although this was the one paper that Einstein himself referred to in that letter to Conrad Habicht as "very revolutionary," in many ways it builds from his other work, and to modern eyes looks less revolutionary than the work on the special theory of relativity. But that is because we have become used to the idea that light exists in the form of tiny particles, called photons. In 1905, that really was a revolutionary concept—although it wasn't exactly new.

Isaac Newton thought of light as a stream of tiny particles, and used this model in his attempts to explain his observations of the way light is bent when it passes through a prism, reflected by mirrors, and can be broken up into all the colors of the rainbow, forming a spectrum. His seventeenth-century Dutch contemporary, Christiaan Huygens, had argued for a different interpretation of the same phenomena, based on the idea that light is a form of wave; but Newton's model held sway (largely because of the God-like status that his successors gave to Newton) until the work of the Englishman Thomas Young and the Frenchman Augustin-Jean Fresnel early in the nineteenth century. Although their contributions were equally important, what became regarded as the definitive proof that light travels in the form of a wave, like ripples on a pond, comes from what has become known as Young's double-slit experiment.

The experiment is based on shining light of a single color

(this would later be interpreted as meaning light of a single wavelength) coming from a light source through two holes in a screen. These could be two parallel thin slits, made with a razor, or two pinholes. We have all seen the pattern of circular ripples that spreads out from a point when a pebble is dropped into still water, and the more complicated pattern of ripples that is produced if two pebbles are dropped into the water simultaneously. The complications in the second pattern are caused by two sets of ripples interacting with one another—as physicists put it, "interfering" with one another. The experiments showed that light spreads out from the holes (or slits) in Young's experiment in just the same way, and that two sets of waves, one from each hole, are interfering with one another. Young (and then many other people) proved this by placing a second screen on the other side of the two slits from the light source, and looking at the pattern of bright and dark stripes made on the second screen. Bright stripes are places where the two sources of light combine with each another to make an extra-high wave, and dark stripes are places where the two sets of ripples cancel each other out, with one going up while the other goes down. This is quite different from the pattern that would be expected if light traveled in the form of tiny particles, like little bullets; indeed, the experiments are so precise that the spacing of the stripes on the second screen can be used to calculate the wavelength of the light involved—proof, indeed, that light travels as a wave.

It's worth putting this dramatic discovery in its historical perspective. Newton had spelled out his ideas about light and color in a great book, *Opticks,* published in 1704. His image of light as a stream of tiny particles held sway for almost exactly a hundred years, until the work of Thomas Young at the beginning of the nineteenth century. That's almost exactly the same as the time interval from Einstein's work in 1905 to the present day—and the interval from Young to Einstein is the same as the span from Newton to Young. To suggest at the beginning of the nineteenth century that Newton had made a major blunder was very much as if evidence were uncovered today showing that Einstein had made a major blunder. The discovery was dramatic, and it took time for people to be convinced. But in the hundred years from Young to Einstein, a great deal more evidence did come in to show that light travels as a wave.

We have already mentioned the most important piece of that evidence—James Clerk Maxwell's discovery of the equations that describe how electromagnetic waves (or "vibrations of the ether," as they would have put it then) move through space. Maxwell's equations describe waves, and they predict the speed with which those waves move. This speed is exactly the same as the speed of light. What more proof could be needed that light travels as a wave? By 1900, the idea that light, and other forms of electromagnetic radiation, existed in the form of waves seemed as solid a foundation of science as the idea that apples fall downward from trees. But then the first crack appeared in this foundation.

Two of the big areas of scientific interest in the middle and late nineteenth century were thermodynamics, which dealt with energy, and light, which had been identified as a wave and was also a form of radiant energy—"light" in this context refers to all kinds of electromagnetic radiation, including invisible infrared heat, radio waves, and ultraviolet light. It was clear that there is a relationship between heat (energy) and light. A piece of iron that is just warm to the touch doesn't radiate any visible light at all, but as it is heated further it glows first red, then orange, then white hot as its temperature increases. Indeed, the relationship between color and temperature was so well known in a qualitative way that in the days before accurate scientific measurements were possible, potters used to gauge the temperature of their kilns by looking at the color of the pots they were firing. But what was the precise, quantitative relationship between light and energy? What were the equations that could describe, or predict, the color of a hot object from its temperature alone?

The first person to tackle this puzzle in a quantitative way was the German physicist Gustav Robert Kirchhoff at the beginning of the 1860s. Kirchhoff was especially interested in spectroscopy, studying the distinctive patterns of lines in a spectrum (looking not unlike a modern bar code) corresponding to different elements. But he also developed a thermodynamic approach to understanding the relationship between light and energy through his idea of a "blackbody." A blackbody would be an object that absorbed entirely all the radiation that fell on it—a perfect absorber. Of course, it was impossible for experimenters to make a perfect blackbody to study in the lab, but Kirchhoff came up with a very close approximation. He devised an experiment involving a closed container painted black inside, with a tiny pinhole as the only opening to its interior. Any radiation that entered through the pinhole would be absorbed, and only a tiny amount of radiation would escape through the pinhole, making it very nearly a perfect blackbody.

But this was only the first step. According to the thermodynamic rules, an object that absorbed all kinds of radiation should also radiate all kinds of radiation, with no complications involving things like spectral lines. If the container were heated, the glowing walls inside would produce light that would bounce around inside and get thoroughly mixed up before escaping from the pinhole, in the form that became known as "cavity radiation," or (more commonly today) "blackbody radiation." According to the thermodynamic principles, this would be a pure form of light, with a color that depended only on the temperature of the container—the blackbody. Crucially, the color of the radiation did *not* depend on what the container was made of.

But the blackbody radiation is not light of a pure single color. It is always a mixture of different colors—that is, different wavelengths—of light. For any particular temperature, however, there is always more energy radiated in one group of wavelengths, with less energy radiated at both longer and shorter wavelengths. As the cavity is heated, the peak intensity of the light shifts from the longer wavelength end of the spectrum (red) through the familiar colors of the rainbow (orange, yellow, green, blue, and so on). So a red-hot piece of iron (or anything else) radiates *mostly* red light, but also some infrared radiation (at longer wavelengths) and some yellow and orange light (at shorter wavelengths).[11]

A graph representing the spectrum of a black body radiating in this way looks like a little hill, with a peak at a particular wavelength corresponding to the temperature of the black body, and slopes rolling down on either side. Even without any understanding of why this should be so, the discovery had immediate practical uses. For example, the shape of the spectrum of the Sun very closely follows this "blackbody curve" for an object with a temperature of about 6,000°C, so astronomers could measure the temperature of the surface of the Sun (and, indeed, of other stars) without ever leaving the Earth. What was needed to complete Kirchhoff's work, though, was to find an equation that described the shape of this hill, and a physical basis for that equation. That proved extremely difficult, and Kirchhoff, who died in 1887, didn't live to see it.

When he died, Kirchhoff was a professor of physics at the University of Berlin. His successor, who wasn't appointed until 1889, was a thirty-one-year-old physicist of the old school, Max Planck. Planck was so conservative, in scientific terms, that he might almost be described as a reactionary. He hated the way ideas involving probability and statistics, rather than the certainty of conventional mathematical equations, were being introduced into physics by people like Boltzmann, and had yet

---

[11] A white-hot object looks white because the peak of its spectrum is in the middle of the rainbow of colors, so it radiates all the colors, which combine to make white light. If it were even hotter, it would look blue, as some stars do.

to be convinced of the reality of atoms. In 1882 he had stated dogmatically that "despite the great success that the atomic theory has so far enjoyed, ultimately it will have to be abandoned in favor of the assumption of continuous matter."[12] What stuck in Planck's throat was the idea that matter could come in discrete lumps, with gaps in between. He much preferred the image of electromagnetic radiation as smooth and continuous waves, and expected that matter would also be found to be smooth and continuous, no matter how successful the kinetic theory and Boltzmann's ideas on thermodynamics might seem to be.

In 1897, though, Sir Joseph John ("J. J.") Thomson, working at the Cavendish laboratory in England, showed that the streams of radiation known as cathode rays were actually made up of tiny, electrically charged particles, which soon came to be known as electrons. Whatever the reality of atoms, there could be no doubt that matter contained these little particles, and since they carry electric charge it seemed clear that there must be some connection between the behavior of electrons in matter and the way matter radiated light. In particular, Maxwell's equations told physicists that a charged particle vibrating to and fro (an electric oscillator) must radiate electromagnetic waves.

By the end of the nineteenth century, though, the problem of how to describe blackbody radiation mathematically had run into a cul-de-sac—or rather, two cul-de-sacs. In the early 1890s, Wilhelm Wien, a lecturer at the University of Berlin, had

---

[12] Quoted by John Heilbron, in his biography of Planck, *The Dilemmas of an Upright Man,* University of California Press, Berkeley, 1986.

come up with a mathematical description of blackbody radiation that produced a graph exactly matching one side of the spectrum, the short-wavelength side of the hill, but was hopelessly wrong at describing the shape of the curve for longer wavelengths. In 1900, the English physicist Lord Rayleigh found another equation, based on different physical assumptions, that predicted a curve which exactly matched the blackbody curve on the long-wavelength side of the hill, but was hopeless when applied to the shorter-wavelength side.[13] This discrepancy clearly showed that there was some fundamental misunderstanding of the nature of blackbody radiation, and caused consternation among physicists. But in the same year, 1900, Planck, who had been studying the problem of blackbody radiation intensively since 1897, came up with something that looked like it might be the answer to the puzzle.

In October 1900, Planck came up with a formula—an equation—that described the entire blackbody curve accurately, smoothing over the join between Wien's law and Rayleigh's law. This was essentially an empirical result, one worked out by trial and error, and it included two constants, one of which became known as Boltzmann's constant and the other of which, given the label h, became known as Planck's constant. Boltzmann's constant usually turns up in connection with the properties of gases, but, unlike the case of the constant c in Maxwell's equations, there was no obvious physical interpretation of the new

---

[13] His equation was later revised by James Jeans, and became known as the Rayleigh-Jeans law.

constant h. Planck presented his results to a meeting of the Berlin Physical Society on October 14, even though he was well aware that he still lacked any physical basis for the equation he had come up with. But he kept beavering away at the problem, and before the end of the year he had found a physical basis for the equation. It wasn't particularly palatable to him, but it was the best he could do.

Planck was working on the assumption that electromagnetic radiation was emitted or absorbed by matter because of the presence of the relatively newly discovered electrons jiggling about inside the matter. In trying to explain the nature of the cavity radiation studied by Kirchhoff and others, he had to think of the walls of the cavity containing a large number of harmonic oscillators, each corresponding to a jiggling electron. Radiation would be radiated and absorbed, reradiated and re-absorbed, re-reradiated and re-reabsorbed, in a repeating process mixing up all the radiation to achieve a state of dynamic equilibrium, with the maximum amount of disorder, before it could escape from the pinhole.

Unfortunately for Planck, this kind of equilibrium was described by the rules of thermodynamics—indeed, it is known as thermodynamic equilibrium—and describing it mathematically involved some of the statistical techniques developed by Boltzmann (which is where his constant comes in). Even more unfortunately, there was no getting away from the fact that electrons were individual particles, not a continuum, no matter how much Planck might like the idea that matter was continuous. The only way to take account of their collective behavior was, once again, to use the statistical techniques developed by Boltzmann and Maxwell in connection with the study of the behavior of large numbers of atoms and molecules. In particular, Planck was forced to use these statistical methods to calculate the property of the array of harmonic oscillators known as entropy. It's worth elaborating a little on this, since it would also be central to Einstein's work.

Entropy is a very important concept in thermodynamics, but it can be understood very simply as a measure of the amount of disorder in a system. The entropy of an isolated system (which means anything left to its own devices, with no constructive input of energy from outside) always increases, which is a scientific way of saying that things wear out. If you build a house and leave it untended for a few hundred years, it will crumble away; but if you put a pile of bricks in a heap on the ground and leave them alone they will never spontaneously arrange themselves into a house. People, and other living things, can hold the increase of entropy at bay for a time by making use of the energy in the food we eat (which ultimately comes from the Sun and is stored in plants); but in the very long term everything wears out.

Disordered systems have more entropy than ordered systems. That's why an ice cube placed in a glass of water melts, evening out the difference between the water and the ice, and why water diffuses the "wrong way" through a semipermeable membrane. It's why the radiation inside one of Kirchhoff's cavities gets scrambled up into a complete mess. A chess board painted in black and white squares is an ordered system with relatively low entropy, but the same piece of board painted with the same amount of black and white paint but mixed to a uniform gray color has less order, and more entropy. If you pour a can of black paint and a can of white paint into a bucket, you don't end up with black on one side of the bucket and white on the other side, buy a gray mixture. This mixture has higher entropy. Similarly, the completely mixed-up light that emerges from the pinhole in the form of cavity radiation has very high entropy. One of the most important features of the statistical approach to thermodynamics is that systems are more likely to be found in high-entropy states than in low-entropy states.

You have to give Planck credit for biting the bullet and using the ideas and techniques he abhorred to work out a physical basis for the equation for blackbody radiation. But what he discovered was startling. According to this interpretation of events, light was not being emitted or absorbed in a continuous fashion, but in the form of little lumps, which he called quanta. Speaking to the Berlin Physical Society again on December 14, 1900, Planck described his new work, and said:

> We therefore regard—and this is the most essential point
> of the entire calculation—energy to be composed of a very
> definite number of equal finite packages, making use for
> that purpose of a natural constant $h = 6.55 \times 10^{-27}$ erg sec.

Don't worry about the units for h, just notice how tiny it is—a decimal point followed by 26 zeroes before you get to the 6. And the "finite packages" are only equal for each color, or wavelength, of light. The size of each packet of energy E is given by the equation $E = h\nu$, where ν is the frequency (proportional to 1 divided by the wavelength) of that particular color of light. Instead of establishing that matter is continuous, on the face of things Planck had found that electromagnetic radiation was *not* continuous! But he didn't interpret his results that way.

Looking back from 1931, Planck described his breakthrough as "an act of desperation" and said the idea of the quantum of energy was:

> A purely formal assumption and I didn't give it much thought, except only that, under all circumstances and at whatever cost, I had to produce a positive result.[14]

He still thought of light as a continuous wave (after all, Young's experiment still worked!), and regarded the quanta as some kind of mathematical tool, only useful in the statistical process of adding up the contributions of all the harmonic oscillators. Nobody else knew what else the quanta could actually be either; but Einstein, who had studied Kirchhoff's work while he was at the ETH, heard news of Planck's work while he was teaching at Winterthur in 1901, and was deeply puzzled. He kept the problem of what Planck's quanta really were in his mind all the time he was developing his own statistical skills, with the work described in his first scientific papers. Since

---

[14] Interview quoted by Fölsing.

nobody else had made any progress with interpreting Planck's equation (indeed, physicists were largely too puzzled even to try to explain it), when he had learned enough to be able to tackle the puzzle properly at the beginning of 1905 Einstein was able to pick up the thread exactly where Planck had left off.

The first great paper Einstein wrote in 1905 (he finished it on March 17) is often referred to as the photoelectric paper, not least because when Einstein eventually got his Nobel Prize the citation referred to that aspect of his work. But the section on the photoelectric effect was only a relatively small part of the paper, one of several examples that Einstein used to illustrate the importance of the concept of light quanta. Nevertheless, it was a very important part of the paper, and these ideas had also been turning over in his mind since 1901.

It all started, as hinted earlier, with the work of Philipp Lenard, a German physicist who carried out a series of experiments, starting in 1899, investigating the way ultraviolet light shining onto the surface of a metal in a vacuum could cause it to emit electrons (then still known as cathode rays; remember that electrons had only been identified as particles in 1897). The immediate impact Lenard's work (of which more shortly) had on Einstein when he learned about it in 1901 can be seen from the beginning of a letter he wrote to Mileva late in May that year:

I just read a wonderful paper by Lenard on the generation of cathode rays by ultraviolet light. Under the influence of this beautiful piece I am filled with such happiness and joy that I must absolutely share some of it with you.[15]

What's especially interesting about that letter is that it is the first one he wrote to Mileva after she informed him that she was pregnant; he gets around to offering his response to that news only later on in the letter. It is quite clear where Einstein's priorities lay. But the important point in terms of the genesis of the "very revolutionary" paper on the light quantum is that this was not something that sprang suddenly into his mind fully formed at the beginning of 1905, but was the project that he had been working on, from time to time, for four years. Einstein was a genius, but he was a methodical and hardworking genius.

In some ways, he was also cautious, as we have seen with his choice of title for the Brownian motion paper. Although he knew his work on the light quantum was revolutionary, he was careful to choose a title for the photoelectric paper that would not be an immediate turnoff to readers of the *Annalen der Physik*. He settled on "On a Heuristic Point of View Concerning the Production and Transformation of Light," thereby giving the impression that all he was offering was a convenient mathematical device for carrying out calculations, not a suggestion that light might really be made up of a stream of particles. But he immediately pulled the rug from under that cozy assumption in the opening paragraphs of the paper.

First, he spelled out the dilemma confronting physics. "While

---

[15]*Collected Papers.*

we consider the state of a body to be completely determined by the positions and velocities of an indeed very large yet finite number of atoms and electrons, we make use of continuous spatial functions to determine the electromagnetic state of a volume of space." Which means that this "electromagnetic state" can never be described by a finite number of quantities, no matter how big that number is. Einstein spelled out that the energy of a material body cannot be broken down into an arbitrarily large number of arbitrarily small parts, but that in contrast, according to Maxwell's equations, the energy of light spreading out from a source gets weaker and weaker indefinitely as it spreads. These two concepts of energy had to come into conflict where matter and light interact with one another, as in the interior of a blackbody cavity, or in the photoelectric effect.

Then Einstein made a key point. Of course, he acknowledged, every traditional optical experiment (such as Young's experiment) produced results that were consistent with the wave model of light. But "one should keep in mind, however, that optical observations refer to time averages rather than instantaneous values." In other words, light could indeed be composed of tiny particles, provided those particles were so small that their combined effects averaged out on the scale of the usual optical experiments to give the appearance of a smooth continuum. Einstein then gave a list of several recent experiments, including studies of blackbody radiation and Lenard's work on the photoelectric effect, which conflicted with the classical view of light as a wave, and warned his readers what was coming:

According to the assumption considered here, in the propagation of a light ray from a point source, the energy is not distributed continuously over ever-increasing volumes of space, but consists of a finite number of energy quanta localized at points of space that move without dividing, and can be absorbed or generated only as complete units.

In other words, if you had infinitely sensitive eyes and looked at a source of light from very far away, you would not see a continuous faint glow, but individual flashes of light, with total darkness in between, as individual light quanta arrived at your eyes.[16]

If that didn't whet the appetite of his potential readers, nothing would. Without more ado, Einstein plunged into the mathematical part of his paper. He began by taking a fresh look at Planck's calculations, correcting some errors in Planck's own work and coming up with a new derivation of the key equation, describing the blackbody curve, in which Planck's constant appears. Then, in a tour de force argument at the heart of his paper (the bit for which he should have got the Nobel Prize), Einstein compares the entropy of a certain volume of monochromatic radiation (a box full of light of a single color) with the entropy of a certain volume of gas. He calculates the entropy of the radiation

---

[16] Incidentally, this is exactly what astronomers do now "see," using sensitive detectors called charge-coupled devices, when they point their telescopes toward the faintest and most distant objects in the universe. They can literally count the photons arriving one by one.

for short-wavelength radiation (in the region where Wien's law applies) and the entropy of an equivalent box of gas from what were by then the standard techniques developed by Boltzmann—and he got the same answer. His conclusion is that:

> Monochromatic radiation of low density (within the range of validity of Wien's radiation formula) behaves thermodynamically as if it consisted of mutually independent energy quanta.

This is arguably the most revolutionary sentence written in science in the twentieth century, given the success of Maxwell's equations. What Einstein is saying is that, as far as thermodynamic properties such as entropy are concerned, a gas behaves as if it is made up of very many tiny particles (atoms and molecules), and electromagnetic radiation behaves as if it is made up of very many tiny particles ("atoms of light," or photons). At this basic level, there is no fundamental difference between matter and light after all. And there is a point—the most important point of all—which might easily be missed by the casual reader. Einstein has reached this conclusion without having to assume anything at all about the way light interacts with the harmonic oscillators that are at the heart of Planck's treatment of the problem. He is explicitly saying that this graininess of light is an intrinsic property of the light itself, not something to do with the way light interacts with matter. He hasn't just reproduced Planck's result, but has done something much more fundamental than Planck ever did.

It is only after dropping this bombshell that Einstein goes on to consider the implications of his discovery for other areas of physics, most notably the photoelectric effect. He did *not*, as many people imagine, come up with the light quantum idea from the photoelectric effect, but used the photoelectric effect to demonstrate the power of this idea. And it still perfectly demonstrates the power of this heuristic point of view concerning the nature of light.

The curious thing that Lenard had discovered, which had made Einstein so excited in May 1901, was that the energy of an electron produced by the photoelectric effect did not depend on the intensity of the light (how bright it was), but it did depend on the wavelength of the light. The effect only happens at all for ultraviolet light, which covers a range of wavelengths even shorter than the wavelength of blue light and cannot be seen by the human eye, so "color" is not really the right word to use; but in a sense the energy of the electrons produced by the photoelectric effect depends on the color of the light shining on the metal surface. Since the energy of the electrons determines their speed, you can also say that the speed with which the electrons are ejected depends only on the color of the light shining on the surface.

Lenard's discovery runs counter to common sense, because a bright light carries more energy than a dim light. You would expect a bright light to knock electrons out of the metal surface more energetically, so they would move away faster. But Lenard found no such effect. If he used ultraviolet light with a particular wavelength, the ejected electrons always escaped with the same speed. If he turned the brightness of the light up, more electrons were ejected, but still with the same speed; if he turned the brightness of the light down, fewer electrons were ejected, but still with the same speed.

This was impossible to explain on the wave model of light. But it was utterly simple to explain using Einstein's heuristic principle. Planck's equation implied that electromagnetic radiation only existed in little packets of energy, quanta with $E = h\nu$. For a particular wavelength (or frequency) of radiation, every quantum has the same energy. So energy could be handed over to electrons only in quanta that size. As Einstein put it:

> The simplest conception is that a light quantum trans-
> fers its entire energy to a single electron.

In which case, as long as ν was the same every electron would receive the same amount of energy and would rush away from the metal surface with the same velocity. If you turned up the brightness of the light, there would be more quanta, but they would still each have the same energy $h\nu$, so there would be more ejected electrons, but each still moving with the same velocity. The only way to make the electrons move faster would be to use a different wavelength of light, with a bigger value of the frequency ν (which means a shorter wavelength, farther into the ultraviolet).

This was very nearly what Lenard had found, but not quite. The trouble was, the experiments involved were still extremely difficult, and although Lenard had found that shorter-wavelength ultraviolet light did produce ejected electrons with more energy, he couldn't measure the energy precisely. The results weren't good enough to say exactly how much extra energy the electrons got for a particular change in wavelength. Einstein's calculations did predict a very precise relationship, but it was more precise than the experimental results. His equation agreed with the experimental data, but within the range of uncertainty allowed by the data it was conceivable (if unlikely) that some other equation might work just as well. So all he could say in 1905 was that:

> As far as I can tell, this conception of the photoelectric effect does not contradict its properties as observed by Mr Lenard.

With no absolute experimental confirmation of the validity of Einstein's calculations, the idea of light quanta (especially "particles" of light that still somehow had wavelengths associated with them) was so shocking to physicists in 1905 that Einstein's paper was largely ignored for years. The only person who really took much notice of it was an American experimental physicist, Robert Millikan, who was so infuriated that when he heard about it he promptly set out to try to prove Einstein was wrong. But it's an indication of how little attention was paid to Einstein's photoelectric paper when it was published that Millikan, who was already working on the photoelectric effect and investigating other properties of electrons in 1905, didn't even learn about the paper for several years.

Millikan worked at the University of Chicago, and was thirty-seven in 1905, eleven years older than Einstein and ten years younger than Planck. But it was only in 1912, when he was forty-four, that he began his determined effort to measure the properties of electrons ejected by the photoelectric effect accurately enough to test Einstein's predictions and—he firmly expected—prove Einstein was wrong. In a classic example of the scientific method at work, the skeptical Millikan actually found that the relationship between the energy of the ejected electrons and the wavelength of the radiation involved exactly matched Einstein's predictions. But even then, he could not at first accept the reasoning behind Einstein's prediction. When he announced the results of four years of intensive research into the problem in 1916, he said:

> The Einstein equation accurately represents the energy of electron emission under irradiation with light [but] the physical theory upon which the equation is based [is] totally unreasonable.

Nevertheless, he admitted that his results, combined with Einstein's equation, provided "the most direct and most striking evidence so far obtained for the reality of Planck's h."[17]

The Nobel Committee were no less cautious when they

---

[17] *Physikalishel Zeitschrift*, volume 17, page 217, 1916.

awarded Einstein the Physics Prize in 1922 (it was actually the 1921 prize, held over for a year; Millikan received the prize in 1923). The citation noted the work of Millikan in proving Einstein's prediction right, but referred only to "the discovery of the law of the photoelectric effect" (in other words, the equation tested by Millikan) and avoided mentioning the physical model on which the equation was based. But as it happened, just a year later, in 1923, new experiments involving the interaction between electromagnetic radiation and electrons finally established the reality of light quanta, which were then given the name photons by the American chemist Gilbert Lewis in 1926.

Although it is not strictly relevant to our story, it's amusing to see how Millikan rewrote his own history of these events as the reality of photons became more and more firmly established. Having entirely dismissed the physical theory on which the equation was based in 1916, in 1949 he wrote in an article in the journal *Reviews of Modern Physics* that "I spent ten years of my life testing that 1905 equation of Einstein's and contrary to all my expectations, I was compelled in 1915 to assert its unambiguous verification in spite of its unreasonableness," while in 1951, two years before he died, he wrote in his autobiography that "I think it is correct to say that the Einstein view of light quanta, shooting through space in the form of localized light pulses, or, as we now call them, photons, had practically no convinced adherents prior to about 1915, by which time convincing experimental proof had been found." No mention here that even Millikan himself had still not been convinced of the reality of light quanta in 1915.

As we have mentioned, part of the problem of convincing

scientists that light quanta were real was the enormous success of the wave model of light, and in particular Maxwell's equations. At the end of the nineteenth century, it seemed quite clear that particles were particles, and waves were waves. When cathode rays were discovered, nobody knew if they were waves or particles until J. J. Thomson devised the experiments which proved that they were particles. Then, they could be neatly labeled and the possibility that they might be waves forgotten. It was equally natural to assume that light could only be one thing or the other. It wasn't until well into the 1920s that physicists began to come to terms with the uncomfortable truth that it was possible for light to somehow be both a wave and a particle, and to realize that the everyday laws of common sense do not apply on the very small scale of entities such as photons and electrons.

This wave-particle duality lies at the heart of quantum physics, and it is now well established that just as light, which was formerly thought of as a wave, behaves like a stream of particles under some circumstances, so electrons (and other entities that were formerly thought of as particles) have a dual nature and behave under some circumstances like waves. Electrons can even be made to interfere with one another in a variation on Young's experiment. Einstein was the first person to understand that light could behave as a wave under some circumstances (as in Young's experiment) and like a particle in other circumstances (as in the photoelectric effect). This flexibility of approach allowed him to keep faith with the aspects of the wave model that worked—notably Maxwell's equations—even while he was rejecting the wave model in situations where it did not apply. With the photoelectric paper submitted for publication in mid-March 1905, Einstein's long fascination with light was about to bear fruit in an even more spectacular way—once he had finished writing up the paper that would become his PhD thesis and his paper on Brownian motion.

The last of the four great papers of Einstein's annus mirabilis emerged from his fertile brain soon after he had submitted his paper on Brownian motion to the *Annalen der Physik*. The breakthrough was triggered, he later recalled, by a discussion with his old friend Michele Besso, sometime in the middle of May.[18] An intense burst of activity over the next six weeks saw the key paper on the special theory of relativity delivered to the *Annalen der Physik* on June 30 (after Mileva had carefully checked Einstein's calculations for slips, a mundane task that didn't even earn her an acknowledgment in the paper) and published at the end of September, in the same week that the editor of the *Annalen der Physik* received a second paper on the subject, in which Einstein spelled out the famous relationship between mass and energy.[19]

---

[18] The genesis of the special theory was described by Einstein in a lecture in Japan, in 1922; the lecture was reprinted in *Physics Today* in August 1982.

[19] Of course, the "special theory" paper was not known by that name at the time; Einstein introduced the name in 1915, to distinguish it from his general theory of relativity. But we will use the name, since as with the "Brownian motion" paper, we have the benefit of hindsight. "Special" in this context means the theory is a "special case" dealing only with objects traveling at constant velocities; the general theory deals with accelerations as well. But it is always "special theory of relativity"; there is no such thing as the "theory of special relativity," since it is the theory that is "special," not the relativity!

Curiously to modern eyes, this key paper about the nature of space and time is actually titled "On the Electrodynamics of Moving Bodies." This reflects the importance of light—an electromagnetic entity—in Einstein's theory, but also highlights the way in which the puzzle of relative motion had developed in the 1890s. Following the success of Maxwell's equations, physicists in the last quarter of the nineteenth century were convinced that light was a form of vibration in the ether, and various experiments were carried out to try to measure the motion of the Earth through the ether. If light travels at a certain fixed speed through the ether, as Maxwell's equations seemed to imply, and the Earth is moving in the same direction, then you would expect from everyday experience of how speeds add up that the speed of that light relative to the Earth would be less than the speed of that light through the ether; conversely, if the Earth were running head-on into a light beam traveling through the ether, then you would expect the speed of the light beam measured in the experiments to be equal to its speed through the ether *plus* the speed of the Earth through the ether. Making such measurements proved extremely difficult (chiefly because the speed of light is so big, 300,000 kilometers per second), but the predictions encouraged experimenters to develop new techniques in the 1880s and 1890s that were accurate enough to measure the calculated effects. But even when these experiments became sophisticated enough to take account of things like the Earth's movement around the Sun, and its daily rotation on its own axis, they always measured the same velocity for light, whether it was moving in the same direction as the Earth, in the opposite direction to the Earth, or at any angle across the line of the Earth's motion.

The first person to take these results seriously and try to find an explanation for what was going on (rather than just assuming the experimenters were making a mistake) was George Fitzgerald, who was professor of natural and experimental philosophy at Trinity College, in Dublin. In 1889 he wrote a paper, which he sent to the American journal *Science,* in which he pointed out that the experimental results could be explained if the experimental apparatus (and everything else) shrank slightly in the direction if its motion through the ether. The experimental apparatus involved is much more complicated than a simple ruler, but in effect he said that if your ruler shrank by a tiny amount, then the time taken for light to whiz past the ruler from one end to the other would be a little less, and you would measure a different speed than if the ruler had not shrunk. "Paper," though, is perhaps too grand a word for Fitzgerald's squib, which contained no mathematical calculations and consisted only of a single paragraph. It was clear that in order for all experiments to always measure the same speed for light, the shrinking had to obey a precise mathematical formula, which Fitzgerald did not spell out. This shows, for example, that in order to make a meter-long ruler shrink to 99 cm (that is, a reduction in length of just 1 percent), it would have to be moving at one-seventh of the speed of light, 43,000 kilometers per second.

Fitzgerald had no detailed physical explanation for why objects should shrink in this way, and his colleagues in Dublin laughed at the idea. Although the paper was published in *Science,* nobody took any notice.[20] So when the Dutch physicist Hendrik Lorentz came up with a similar idea in 1892, and the appropriate mathematical equation to describe the shrinking effect, he didn't know about the similarity to Fitzgerald's earlier work until this was pointed out by the British physicist Sir Oliver Lodge.

The contraction formula became known as the Lorentz-Fitzgerald contraction, which seems a little unfair both in terms of the chronology and alphabetically. There was, though, a sound reason why it became known as Lorentz-Fitzgerald contraction, not Fitzgerald-Lorentz contraction. Unlike Fitzgerald, Lorentz developed this mathematical equation alongside a physical picture of what might be going on to make moving objects shrink.

In the early 1890s, Lorentz, who was born in 1853, was professor of theoretical physics at the University of Leyden. He had developed a theory of electrodynamics which he was already calling the "electron theory"—although, confusingly from our point of view, this did not involve the particles now known as electrons. He suggested that all matter is made up of electrically charged particles, some with positive charge and some with negative charge, held together by electromagnetic forces. In 1892 he simply referred to these entities as "charged particles," but in 1895 he referred to them as "ions."[21] It was only in

---

[20] Indeed, for some time after his paper had been published on the other side of the Atlantic, Fitzgerald himself didn't know that it had appeared in print.

[21] Again, with a different meaning from what "ion" means to a scientist today.

1899, two years after the identification of cathode rays as streams of negatively charged particles, that he started calling these particles "electrons." The name stuck, even though the "electron theory" did not last.

Like Fitzgerald, Lorentz argued that the experimental observations of the constancy of the speed of light could be explained if moving objects shrank in the direction they were moving, and he came up with the appropriate formula for the contraction. Also like Fitzgerald, he assumed that the cause was the motion of the object *relative to the ether,* which provided a standard frame of reference against which all motion could, in principle, be measured. But he went further by suggesting that the reason why objects shrank in this way was because of a physical effect of the ether on the moving objects. Specifically, he suggested that there was an electric force that was caused by the motion, which had the effect of squeezing the charged particles the moving object was made of. Lorentz took these ideas much further than Fitzgerald (not least because Fitzgerald died in 1901, at the early age of 49), and came up with a complete theory that he published in a Dutch journal, which was not very widely read, in 1904. As well as describing how the length of a moving object was related to its motion relative to the ether, this work raised the ideas of relative time and the synchronization of clocks by using light signals—which was also, as we shall see, a central feature of Einstein's work.

In all of this, Lorentz was encouraged by French mathematician Henri Poincaré, a year younger than Lorentz, who publicized the ideas and became interested in the mathematical foundations of the equations involved. Indeed, the first appearance of the term "relativity principle" was in a lecture Poincaré gave at the World Exhibition in St. Louis in 1904. It was also Poincaré who first used the term "Lorentz transformations" to describe the whole package of equations that Lorentz had derived in his 1904 paper.

Einstein had followed at least some of these developments, and was well aware of the fact that all experiments showed no effect of the motion of the Earth on the measured speed of light, even though he does not seem to have been particularly familiar with the details of all the experiments. He later said that he had spent seven years puzzling over the electrodynamics of moving bodies before the breakthrough in 1905, and this is borne out by a letter he wrote to Mileva in August 1899, in which he said:

> I'm more and more convinced that the electrodynamics of moving bodies as it is presented today doesn't correspond to reality, and that it will be possible to present it in a simpler way. The introduction of the term "ether" into theories of electricity has led to the conception of a medium whose motion can be described without, I believe, being able to ascribe physical meaning to it.[22]

This is where Einstein would make his dramatic breakthrough in 1905. He did away with the ether. Instead of saying that what matters is motion relative to the ether, he said that

---

[22] *Collected Papers.*

what matters is how two objects move *relative to each other*, and that there is no absolute standard of rest against which motion can be measured.

There's another especially intriguing feature of the paper on what became known as the special theory. It contains no references at all to any earlier work, not even that of Lorentz, or the experiments involving the speed of light. Instead, it starts from first principles and Maxwell's equations to build a logical, consistent mathematical structure that leads inevitably to the conclusions about the nature of space and time. By structuring his paper in this way, Einstein is clearly proclaiming to the world that he has discovered a fundamental, absolute truth about the nature of the universe, to rank with such fundamental mathematical truths as the Pythagorean theorem concerning the lengths of the sides of right-angled triangles. It does not depend on experiments or theoretical models, it is part of the very fabric of the way the world works.[23]

In this spirit, Einstein starts with just two facts about the world, what the mathematicians would call postulates or axioms, and constructs the whole edifice of his theory by building upward from those foundations. The first postulate comes straight from the world of the practical application of electromagnetism in dynamos and electric motors, the industry where

---

[23] Shortly before he died, Einstein told his biographer Carl Seelig that in 1905 he knew about Lorentz's work of 1895, but not about his later work or Poincaré's contributions. This may be an exaggeration, but he probably had not actually read Lorentz's 1904 paper, since the *Proceedings* of the Amsterdam Academy were not exactly easy to get ahold of in Bern.

Einstein's father had worked for so long. The nineteenth-century boom in this industry was based on the work of Michael Faraday, an English chemist and physicist, who discovered in 1831 that when a conducting wire moves in a magnetic field, an electric current flows in the wire. Or rather, using modern terminology, he found that when a wire moves *relative to* a magnetic field an electric current flows in the wire. It doesn't matter whether the magnet is fixed in place in the laboratory and the wire moves past it, or whether the wire is fixed in place in the laboratory and the magnet moves past it. Either way, an electric current flows in the wire. As Einstein put it in the opening paragraph of his paper on the special theory:

The observable phenomenon here depends only on the relative motion of conductor and magnet.

And this leaves no role for the ether, since the observed phenomenon is not affected by the motion of either the magnet or the wire relative to the ether. If, for example, both the wire and the magnet move alongside each other in the same direction (any direction) and at the same speed (any speed), there is no current in the wire.

This brings us to the idea of reference frames. A reference frame is the place you make measurements from, like the laboratory in the previous example. We know that the lab is actually being carried along with the Earth's motion, but we can treat it as if it were at rest. All the laws of mechanics (Newton's laws) work perfectly in the lab. Those same laws also work perfectly in a frame of reference moving at a constant velocity relative to the lab—for example, in an aircraft flying at a smooth and steady speed, or, in the example Einstein always favored, in a train rolling smoothly along a track at a constant speed in a straight line (that is, at constant velocity). Such frames, in which Newton's laws work perfectly, are now called inertial frames.

The first person to spell most of this out, long before the days of trains and planes, was Galileo Galilei, in the seventeenth century; his insights were part of the foundations of Newton's work. As far as the physical behavior of matter and Newton's laws are concerned, by 1905 it had been known for hundreds of years that there is no difference in how things behave if your frame of reference is a lab in an immovable building or another lab in another frame of reference moving steadily at a constant velocity relative to the first lab. Newton himself believed that there must be an absolute "standard of rest" in the universe against which all motion could be measured, and this tied in with the idea of the ether. But no experiment involving Newton's laws could ever detect motion relative to this hypothetical absolute rest frame.

Now, in 1905, Einstein had realized that no experiment involving Maxwell's equations could ever detect motion relative to this hypothetical absolute rest frame, either. So there was no need for the ether. Referring to the way electricity is generated by the relative motion of a wire and a magnet, he continued:

Examples of this sort, together with the unsuccessful attempts to detect a motion of the earth relative to the "light medium," lead to the conjecture that not only the phenomena of mechanics but also those of electrodynamics have no properties that correspond to the concept of absolute rest. Rather, the same laws of mechanics and optics will be valid for all coordinate systems in which the equations of mechanics hold.

In this context, "coordinate system" means the same as "reference frame." Einstein is saying that *all* of the laws of physics, both for mechanics and electrodynamics, are the same in any reference frame moving at constant velocity relative to any other reference frame in which those laws apply. There is no special reference frame that can be regarded as at rest in an absolute sense. He called this his first postulate, which, echoing Poincaré, he named the "principle of relativity."

The second postulate is even simpler, and comes straight from Maxwell's equations:

Light always propagates in empty space with a definite velocity V that is independent of the state of motion of the emitting body.

Einstein based this postulate on Maxwell's equations, not on the experiments that tried to detect the effect of the Earth's motion on the measured speed of light. As he realized, all that could be said about those experiments was that they had "failed to detect" any movement of the Earth relative to the hypothetical ether, and failed to measure any change in the speed of light. But there could still be effects too small for the experimenters to have measured. The postulate, though, was precise and unequivocal—as were the results he obtained from these two disarmingly simple assumptions.

The other key ingredient of the special theory of relativity is a direct consequence of the second postulate. For objects moving at constant velocity relative to one another, each object can be regarded as carrying its own reference frame (coordinate system) along with it. Einstein's theory had to deal with the relationship between coordinate systems, clocks, and electromagnetic processes. And the first thing he had to do to develop this theory was to spell out the way the second postulate affects our ideas about time.

In everyday life, we all have an idea in our heads that it is the same "time" everywhere at once. But what does this really mean? If some master clock in London sends out a time signal by radio at noon, and I have a clock which receives that radio signal and automatically sets itself to noon, my clock will actually be a fraction of a second behind the clock in London, because it takes radio waves traveling at the speed of light a certain amount of time to reach my clock. As long as I don't move my clock, and I know how far it is to London, I can get around this problem by building in an allowance for the time taken by the radio waves to travel to the clock. Since light travels at a constant speed, the only completely accurate way to do this is to send a light signal to London and back, time how long the return journey takes, and divide by two to work out the difference. But what if this process is watched by a moving observer?

The first postulate says that this observer is entitled to regard himself as stationary, with the clocks moving past him at a constant velocity. He will see the calibrating light pulse from the clock head off to London and bounce back, but the return journey will be a different distance in his frame of reference because the whole moving coordinate system will have shifted forward while the light was on its journey. So the "stationary" observer and the "moving" observer do not agree on the distance between the clock and London, and they do not agree on what "the same time" means. In other words, measurements of both time and space are relative—they depend on how the person who makes them (the observer) is moving, relative to the things they are measuring. Remember, also, that the second postulate says that each observer will see *all* light pulses, whether they originate in his own frame of reference or the other one, moving at the same speed. Putting the appropriate mathematics into all of this led Einstein to find a system of equations that could be used to transform the measurements made by one observer into the equivalent measurements in any other coordinate frame moving at constant velocity relative to that observer.[24] These coordinate transformations are exactly the same as the equations found by Lorentz a year earlier, although Einstein had not read Lorentz's 1904 paper in 1905. But there is a huge difference in the interpretation of those equations.

---

[24]We do not have space to go through the argument in detail here; the easiest way to understand what is going on is provided by Lewis Epstein in his book *Relativity Visualized.*

Lorentz had found a system of equations that worked, but which were solely based on the need to explain the experimental failure to measure the motion of the Earth relative to the ether. But he still thought in terms of the ether, and there was no underlying principle shoring up his equations. This is a little like the way Planck came up with an equation for the blackbody curve that fitted the curve but had no foundation in terms of an underlying principle. By contrast, as he had with the blackbody radiation, Einstein started out from first principles and proved that the world really must work in accordance with the equations.

The best example of how different Einstein's view of the world was even to that of Lorentz and Poincaré is that he understood the importance of the symmetry in what we still call the Lorentz transformations. Poincaré had noticed that the transformations are symmetrical. If one observer sees a moving object shrunk in the direction of its motion, then the symmetry implies that an observer riding with the moving object sees his own frame of reference as perfectly normal, but the first observer and everything else in his coordinate system shrunk. Poincaré dismissed this as a quirk of the equations, with no physical significance—it made no sense to him to turn his worldview around and think of the ether being "shrunk" as it moved past us. But Einstein, who had no need of the ether, saw the symmetry in the transformations as a fundamental truth of profound physical significance.

Einstein proved that an observer in one inertial frame would perceive objects in a different inertial frame shrunk in the direction of their motion relative to him, and he would see clocks in the other inertial frame running more slowly than clocks in his own inertial frame. An observer in the other inertial frame would see the mirror image of this—he would see the first observer's clocks running slow, and the first observer's rulers and other equipment (and, indeed, the first observer) shrunk. All of this has now been confirmed by experiments. The effects are very small unless the relative velocities involved are a sizable fraction of the speed of light (which is why we don't notice them in everyday life and they are not common sense), but particles are routinely observed traveling at such speeds in atom-smashing machines like those at CERN, in Geneva, and Fermilab, in Chicago. The special theory of relativity has been proved accurate time and again. It also makes one further prediction, which Einstein himself didn't spot in June 1905 but quickly realized and wrote about in another paper that was a kind of footnote to the paper on the special theory. That realization led to the most famous equation in science—although, disappointingly, the equation does not appear in its familiar form in the paper itself.

In the summer of 1905, Einstein wrote a letter to Conrad Habicht in which he said:

One more consequence of the paper on electrodynamics has also occurred to me. The principle of relativity, in conjunction with Maxwell's equations, requires that mass be a direct measure of the energy contained in a body; light carries mass with it. A noticeable decrease in mass should occur in the case of radium. The argument is amusing and seductive; but for all I know the Lord might be laughing over it and leading me around by the nose.[25]

Einstein mentions radium because this archetypal radioactive element emits energy in the form of radiation and heat all the time. The origin of this energy had been a puzzle for science ever since the discovery of radium by Marie and Pierre Curie at the end of the 1890s; Einstein's discovery implied that the matter radium was made of was slowly being converted into energy, so that the radium itself would gradually lose mass. There had previously been suggestions that electromagnetic energy might be associated with mass, and even that the electron's mass might be completely attributed to its electromagnetic field. But Einstein was suggesting something different, that *all* matter had an energy equivalent, an energy that might in principle be liberated; and he calculated a precise value for this energy.

His little paper pointing this out (just three pages long in its printed form) was received by the *Annalen der Physik* on September 27. It was published before the end of the year but in the next volume of the *Annalen* (volume 18), not the one containing the three great papers on Brownian motion, light quanta, and the special theory of relativity (volume 17, now a valuable collector's item). Still using *V* to denote the velocity of light, in his own words Einstein concluded:

---

[25] *Collected Papers.*

If a body emits the energy L in the form of radiation, its mass decreases by L/V^2. Here it is obviously inessential that the energy taken from the body turns into radiant energy, so we are led to the more general conclusion:
The mass of a body is a measure of its energy content; if the energy changes by L, the mass changes in the same sense by $L/9 \times 10^{20}$ if the energy is measured in ergs and the mass in grams.

In those units, the speed of light is 3×10^{10} cm per second, and 9×10^{20} is the speed of light squared. Putting E for the energy rather than L, c for the speed of light rather than V, m for mass, and rearranging the equation slightly to have the energy, rather than the mass, on the left-hand side, the equation Einstein discovered becomes

$$E = mc^2$$

This is the only equation that everybody knows.

Like all of the results from the papers written by Einstein in his annus mirabilis, this prediction has since been amply confirmed by experiment—not least the awesome "experiment" of the nuclear bomb. It is now understood that the conversion of mass into energy provides the energy source that keeps the Sun and stars shining, and is therefore the ultimate source of the energy on which life on Earth depends, which makes this little paper just about the most important footnote in scientific history.

As we hope we have made clear, however, all of the work that Einstein produced in 1905 was of its time. The statistical ideas that underpinned the doctoral thesis and the work on Brownian motion were a significant step forward, but still part of the mainstream of the investigation of atoms and molecules; the light quantum paper jumped off from the work of Max Planck, and also used statistical ideas from thermodynamics; even the paper on the special theory, although strikingly different in its foundations from the approach used by Lorentz and Poincaré, came up with the same transformation equations, and it is not unlikely that something similar to the special theory would soon have emerged from the line of thinking pioneered by Poincaré himself. In isolation, each contribution was something that an individual physicist at the height of his powers might have been proud of, as the biggest achievement of his career—even the relatively mundane PhD paper, since to many physicists a PhD is the greatest achievement of their academic lives.

What made Einstein so special, and the annus mirabilis so miraculous, was that all four pieces of work were produced by the same young man, working outside the mainstream of scientific life, in his spare time, while holding down a demanding job at the patent office that required his attendance there six days a week for eight hours a day. There was also the inevitable disturbance caused by a year-old baby back at his apartment. And it wasn't as if it really took him a year to do all this. Although, admittedly, a lot of prior thought had gone into all this work, the four great papers were actually written between March and

June 1905, at a rate of one a month, and the $E = mc^2$ paper was finished just three months later, at the end of September.

It would take a little while for the importance of all this to sink in, and for the true genius of Einstein to be appreciated. Indeed, he even stayed at the patent office (partly by choice) for another four years before at last becoming a university professor at the age of thirty. He would also continue to make major contributions to physics until he was in his forties, a remarkably long time for any theoretical physicist. But the rest of his life, and his place in history, would be forever colored by that outburst of creativity in 1905.

DR. HANS EINSTEIN—A BABY DURING THE ANNUS
MIRABILIS—IN PRINCETON FOR HIS FATHER'S
FUNERAL. SHOWN WITH DR. OTTO NATHAN, EXECUTOR
OF THE EINSTEIN ESTATE.

CHAPTER THREE

The Last Fifty Years

The work Einstein published in his *annus mirabilis* didn't immediately set the scientific world on fire, but it was noticed, and drew him into correspondence with a widening circle of physicists, many of whom were astonished to find that he was a junior patent officer and not a professor at the University of Bern. In May 1906, he wrote in a letter to Lenard that "my papers are meeting with much acknowledgment and are giving rise to further investigations. Professor Planck (Berlin) wrote me about it recently."[1] The paper Planck was particularly interested in was not, however, the one on light quanta,

---

[1] Quoted by Fölsing.

about which he had reservations, but the special theory of relativity, of which he was an early and enthusiastic champion.

Einstein himself continued to publish, though not at the prodigious and unsustainable rate of 1905, and his first paper to really make waves in the scientific community appeared in 1907. In this work, Einstein used the idea of energy quanta, combined with his now familiar statistical approach, to explain the way the temperature of an object increases as it absorbs heat. The kinetic theory explains in a qualitative way that the rise in temperature of a solid body as it absorbs energy means that the atoms and molecules vibrate more strongly when the material is hotter. Einstein introduced the idea that the energy being absorbed by the individual atoms and molecules can only be accepted in quanta with energy $h\nu$. He was able to explain otherwise puzzling features of the process and found a formula for the specific heat of a body, which is a measure of how much its temperature rises when a certain amount of heat is absorbed. Planck's colleague in Berlin, the professor of physical chemistry Walther Nernst, took up the idea and incorporated it into his own work on thermodynamics and specific heat, which established over the next few years that it was essential to incorporate quantum ideas into any satisfactory understanding of the thermodynamic behavior of solid objects.

But the most important outside contribution, not only to popularizing Einstein's ideas but (eventually) to shaping the way Einstein's own work would develop, came in September 1908, from Hermann Minkowski, formerly a professor of mathematics at the ETH who had been one of Einstein's teachers, but was now based at the University of Göttingen. Minkowski had been fascinated by the special theory as soon as he saw Einstein's paper, and astounded that a pupil he remembered as a "lazy dog"[2] should have come up with something so profound. He accepted Einstein's ideas without question, and set about reformulating them into a more elegant mathematical package. What he found was that everything contained in the special theory of relativity could be described in terms of geometry—provided that time was regarded as a fourth dimension, on a par with the three familiar dimensions of space.

The natural way to get a handle on this is to think of some everyday experiences in terms of geometrical coordinates. If you draw a triangle on a piece of graph paper, you can specify the triangle completely by giving the coordinates on the grid of the three corners of the triangle, then drawing straight lines to join them up. Similarly, if you arrange to meet someone on the corner of, say, Fourth Street and Main, you are specifying a geometrical location in terms of similar coordinates in two dimensions. If you arrange to meet in the coffee shop on the second floor of the building at the corner of Fourth and Main, you are specifying the location in three dimensions, since the level in the building now comes in as an extra coordinate. And if you say you will meet in the coffee shop on the second floor of the building on the corner of Fourth and Main at three o'clock, you have introduced a fourth coordinate, the time.

Minkowski showed how the mathematics behind things

---

[2] Comment made to Max Born; see, for example, Fölsing.

like shrinking rulers and clocks that run slow could be incorporated into this kind of language, set in a framework of a four-dimensional entity, which became known as space-time. Crucially, this means that there are properties of objects which always stay the same in space-time, even if they look different to different inertial observers in three dimensions. A nice analogy is with the length of the shadow cast by a pencil on a wall. By twisting the pencil about in three dimensions you can make the two-dimensional shadow longer or shorter, but the pencil always stays the same length. In four dimensions, the equivalent property to length is called extension, and by twisting objects around in four dimensions you can change the length of the "shadow" it casts in three dimensions. The fact that time stretches while space shrinks for a moving object reflects the fact that the four-dimensional extension in space-time stays the same—in a sense, the two effects balance each other.

Minkowski presented his ideas in a lecture given in Cologne at the beginning of September 1908. Introducing his talk, he said:

> The views of space and time which I wish to lay before you have sprung from the soil of experimental physics, and therein lies their strength. They are radical. Henceforth space by itself, and time by itself, are doomed to fade away into mere shadows, and only a kind of union of the two will preserve an independent reality.[3]

---

[3] See Pais.

Minkowski never lived to see those words in print, since he died from complications resulting from appendicitis in January 1909, when he was just forty-four years old. But it is no coincidence that widespread acceptance of Einstein's ideas, broad recognition of his abilities, and the opening up of doors into academic life all followed after Minkowski's reformulation of the special theory.

At first, Einstein himself was slightly miffed about the way the mathematicians had picked up his ball and run off with it. Perhaps not entirely seriously, but with an undercurrent of irritation, he commented that "since the mathematicians have attacked the relativity theory, I myself no longer understand it," and said "the people in Göttingen sometimes strike me not as if they wanted to help one formulate something clearly, but as if they wanted only to show us physicists how much brighter they are than we."[4] But he soon changed his tune, when he discovered that Minkowski's geometrization of the special theory was one of two important steps that would eventually put him on the road to his greatest triumph, the general theory of relativity. The other key step was an insight that struck him in 1907, while sitting at his desk in the patent office. In the lecture he gave in Japan in 1922, Einstein said:

> I was sitting in a chair in the patent office in Bern when all of a sudden a thought occurred to me: "If a person falls freely he will not feel his own weight." I was startled.

[4] See Clark.

> This simple insight made a deep impression on me. It impelled me toward a theory of gravitation.

But that theory would take Einstein almost another decade of intermittent but often intense struggle to achieve, even with the help of Minkowski's geometrization of the special theory.

The reason why the "sudden thought" was so important is that as soon as he had completed the special theory Einstein began to attempt to find ways to make the theory more general (hence the name) by adapting it to deal with accelerated motion, not just motion at constant velocities. A freely falling object is accelerating, and Einstein's insight was the realization that this acceleration exactly cancels out the weight of the object. Turning this around (and remember, Einstein was working on this before the time of space rockets, or even high-speed elevators), if you were standing on a platform that was being accelerated in a straight line through space you would feel as if you were standing still and held down by your own weight in a gravitational field. In 1907, Einstein realized that acceleration and gravity are exactly equivalent to one another, so that his general theory, when he found it, would be a theory of gravity, not just a theory of motion. But the path from the insight to the theory was long and tortuous, and Einstein's personal life underwent many changes along the way.

The first change came when he moved on from the patent office in 1909, four years after the annus mirabilis. Until then, his home life hadn't really changed much. On the strength of his doctorate, in 1906 he had been promoted to technical expert

(II class), with an increased salary of 4,500 francs a year, but this made little difference to his frugal lifestyle. In 1908, he became a "privatdozent" at the University of Bern—a kind of part-time unpaid lecturer. This was rather pointless in itself, but an essential step before he would be considered for a full university post. Even before making that next step, in July 1909 Einstein was awarded his first honorary degree (by the University of Geneva), and in the same month he learned that his application for the post of professor of physics at the University of Zurich had been successful. The pay was exactly the same as that of a technical expert (II class) at the patent office, but from taking up the appointment in the autumn of 1909 he could at last call himself Herr Professor Einstein.

After waiting so long for an academic position, over the next few years Einstein moved from post to post in an almost frenetic manner, as his reputation grew and new opportunities opened up. In 1911, he moved to Prague, but returned to Zurich the following year as a professor at the ETH, where he had been a student. This could have been a job for life, but in 1914 he was seduced away by the offer of a job he couldn't refuse in Berlin—the package, put together by Planck and Nernst, included a specially created professorship that need involve no teaching at all, if Einstein so wished it, together with membership in the prestigious Prussian Academy of Sciences, an enhanced salary, and no administrative duties. The only problem, by then, was Mileva.

As Einstein's reputation grew, Mileva found herself increasingly left out of his life. When he wasn't working, he was talking about physics with his new friends, many of them young men fired with enthusiasm for their subject who must have reminded her all too painfully of her own failed ambitions in science. She had enjoyed their time in Bern when Albert was an unknown scientist, but now he was being taken away from her. The first move to Zurich, though, must have seemed like an opportunity to revisit the scene of their own youth, and perhaps make a fresh start. The birth of their second son, Eduard, on July 28, 1910, might have helped, but once again Mileva, now in her mid-thirties, suffered a difficult birth. Eduard turned out to be a sickly baby who needed a lot of attention, while six-year-old Hans Albert could hardly be ignored. Einstein increasingly left all that domestic side of things to Mileva, and then took the family off to Prague.

The move was a mistake. In the last years of the Austro-Hungarian Empire, Prague was a city where the German minority regarded themselves as socially superior to the Czech majority, the Czechs hated the Germans, and both communities disliked the Jews. Strange, gypsylike women from the Balkans were almost beneath contempt, but might just be tolerated if married to a respectable professor. In a letter to Besso, written soon after he arrived in Prague, Einstein said, "My position and my institute give me much joy, only the people are so alien to me."[5] They were even more alien to Mileva, but the situation was saved by the offer of a full professorship at the ETH, back in Zurich for what would turn out to be the last time.

---

[5] *Collected Papers.*

Having just escaped from Prague back to a country and a city she felt comfortable in, it is easy to imagine Mileva's reaction when Albert told her about the job in Berlin. It was the last straw, and although Mileva did travel to Berlin with her husband in the spring of 1914, it wasn't long before she returned to Switzerland, taking her sons with her. So Einstein was alone in Berlin, starting a new life, when the First World War broke out. He had complete freedom to work as he wished, but nobody, at first, to look after the domestic side of things. This became increasingly important as the Allied blockade began its attempt to starve Germany into submission. It was in these circumstances, which became increasingly difficult as the war progressed, that Einstein completed his masterwork, the general theory of relativity. And he only survived the war thanks to the support he soon received from his cousin Elsa, a divorced woman with two daughters who lived in an apartment near his.

Einstein worked obsessively, slept only when he was exhausted, forgot to eat, and neglected personal hygiene. The result was that by 1916 he completed his general theory and presented it to the Prussian Academy, but at the beginning of 1917 he suffered a physical and mental collapse. He did not fully recover for several years, and was nursed back to health by Elsa and her daughters, initially through the worst months of 1917 when the blockade was at its most effective and food was tightly rationed. The relationship was exactly what both of them needed—Einstein, who was forty in 1919, needed someone to look after him; Elsa, three years his senior, needed someone to look after. In 1919 (ironically, on Valentine's Day, February 14)

Albert was divorced from Mileva; on June 2 of the same year he married Elsa.[6] Famously, in the divorce settlement Mileva was promised the money from Einstein's Nobel Prize, even though he had not yet received it; by 1919, it was obvious to all the scientific community that it could only be a matter of time.

So what was the scientific achievement that brought Einstein to death's door? The general theory of relativity is a complete description of the relationship between space, time, and matter on the large scale—which means everywhere that quantum physics does not apply. It deals with all kinds of motion, accelerated or nonaccelerated, and gravity. It describes the fall of an apple from a tree, how light gets bent when it passes near the Sun, and things Einstein hadn't imagined in 1916, such as the expanding universe and black holes. In its full form the theory involves sophisticated mathematics and equations that are daunting to the uninitiated. But its essence can be gleaned quite easily using Minkowski's ideas about space-time.

When we mentioned using coordinates to describe a triangle on a sheet of graph paper, we took it for granted you would be thinking of a flat sheet of paper. Einstein's special theory of relativity describes the way things move about in what is called flat space-time. But you can also imagine triangles drawn on the surface of curved objects, such as a sphere. In that case, the laws of geometry are different from the ones that apply on a flat surface. For example, on a sphere the angles of a triangle don't

---

[6] Einstein's terminally ill mother, Pauline, was moved to Berlin in December that year to spend her last months there. She died in February 1920.

add up to 180 degrees, and lines that start out parallel to one another (like the lines of longitude crossing the equator) can end up crossing each other (in this case, at the north and south poles). Einstein's general theory of relativity describes how things move in curved space-time, and the curvature in space-time is caused by the presence of matter.

The usual example is to think of a stretched rubber sheet, like a trampoline. This is flat, and if you roll marbles across it they travel in straight lines. Now imagine dumping a heavy weight, like a bowling ball, on the stretched sheet. It makes a dent—it curves "space-time"—and if you roll marbles across the curved surface they will follow curved paths past the bowling ball.

Among other things, the general theory of relativity predicts exactly how much space-time will be curved in this way near a massive object like the Sun, and how light rays (which we are used to thinking of as traveling in straight lines) follow curved paths near the Sun as a result. This bending of light rays would show up from Earth, if we could look past the Sun at stars beyond the Sun, as a tiny sideways shift in the apparent positions of those stars in the sky—*if* the light from those stars wasn't overwhelmed by the brightness of the Sun. Einstein realized that the stars behind the Sun would be visible during a total eclipse, so the effect could be measured. In 1919, after the fighting in Europe had stopped but before the formal peace treaty ending the war had been signed, a British expedition to observe a solar eclipse visible from an island off the west coast of Africa confirmed this prediction from a German scientist.

The news, announced in November 1919, that Isaac Newton's theory of gravity had been superseded, that space was curved and that light could be bent as it passed by the Sun, made headlines around the world and made Einstein famous. From now on, he would be the iconic popular image of a great scientist, and his life would never be entirely his own. But the man who was to become the archetypal scientific genius in the public mind was still recovering from illness. His once black hair was turning gray, and he looked older than his forty years. The popular image would bear little resemblance to the dark, handsome, slightly chubby young man who had started to set the scientific world on fire in 1905.

One result of Einstein's newfound fame was that his scientific work began to take a backseat. Besides, few scientists achieve anything significant in the way of original research after their fortieth birthday. But he had already made another contribution of great significance, after completing his general theory of relativity and before his breakdown, which would have been a highlight of anyone else's career. In 1916, Einstein had returned to the puzzle of how light interacts with matter, armed now with the model of the atom as a tiny central nucleus surrounded by a cloud of electrons developed by Danish physicist Niels Bohr from the experimental work of New Zealand physicist Ernest Rutherford. Using the idea that electrons "jump" from one energy level to another inside the atom as they emit or absorb light quanta (photons), he discovered how suitably energized atoms could, in principle, be made to release a pulse of light quanta all with the same wavelength, at the same time, as an energetic beam of pure light. This "stimulated emission of radiation" is the idea behind the laser, although the technology took decades to catch up with Einstein's insight.

Einstein would make just one more important contribution to the quantum theory, in 1924 when he was forty-five years old. But between 1919 and 1924, although he continued to carry out what was by his standards routine physics, he enjoyed some of the trappings of his fame, traveling and (especially after the award of the Nobel Prize) beginning to revel in the role of a kind of father figure to the rising generation of physicists.

Although economic conditions in Berlin were terrible after the war and there was political chaos, the presence of colleagues such as Planck to discuss physics with encouraged Einstein to keep his base there, even though he was offered a special professorship back in Zurich, in the comfortable safe haven of Switzerland. But that did not stop him traveling abroad at almost every opportunity when invited—he even agreed to visit Zurich to give a regular series of lectures each year, not least since payment in the stable Swiss currency would go a long way to ease his living conditions in Berlin. In 1921 he made his first visit to the United States, a lecture tour as part of a fund-raising program for the Hebrew University in Jerusalem. His reception by the public resembled that of a modern pop star, and his lectures on relativity theory drew packed houses, while he was also awarded civic receptions and honorary degrees.

Considering his global status by 1921, it might seem that the Nobel Committee was rather tardy in awarding Einstein its physics prize. In fact, he had been nominated every year since 1910 except 1911 and 1915. Most of the people who did receive the award in those years thoroughly deserved it, although the curious exception is Nils Dalén, a Swedish inventor given the 1912 prize for his automatic regulator to control gas-fueled lighthouse lamps. The same year, Einstein was nominated unsuccessfully for the special theory of relativity. And in 1916, when Einstein was nominated for his work in molecular physics, no award was made. In 1921, the award was deferred, and then awarded to Einstein the following year (specifically for his work on the photoelectric effect), while Niels Bohr received the 1922 prize.

At the time the award was announced, the Einsteins were in Japan. He was there on a lecture tour as a guest of the publishing house Kaizosha, following a successful tour by Bertrand Russell the previous year. Before Russell left Japan, his hosts asked him to name the three most important living people so that they could be invited to follow in his footsteps; Russell gave them just two names, Lenin and Einstein. Since Lenin was busy at the time, they invited Einstein. The invitation was particularly welcome, as it meant Einstein would be away from Germany for at least six months, from October 1922, at a time when the political situation was deteriorating following the assassination by right-wing extremists of Walther Rathenau, a Jew who was the minister of reconstruction. This was part of a pattern of growing anti-Semitism and violence.

By the time the Einsteins returned to Europe in 1923, his fame had reached even greater heights, thanks to the award of the Nobel Prize, but Germany was plunging deeper into the depths of economic collapse, runaway inflation, and violence, all of which paved the way for the rise of Nazism. Amid this turmoil it is doubly surprising that at the relatively grand old age (for a theoretical physicist) of forty-five Einstein would still produce one last piece of really significant science. But he did not do it alone.

Einstein's fame and status in the scientific community meant that he was sent streams of communications from aspiring scientists, as well as from his established colleagues, from all over the world. In June 1924 he received a letter from a young Indian physicist, Satyendra Bose, who was based in a city then known as Dacca.[7] The letter accompanied an unpublished scientific paper in which Bose developed Einstein's ideas on light and radiation in new ways. Bose had found a new way to derive the equation for blackbody radiation, without assuming that light behaved as a wave at all but simply as a quantum gas obeying a new kind of statistical law. Einstein had cut his scientific teeth on statistical laws, and immediately saw the importance of this discovery. He first translated Bose's paper (which was written in English) into German and got it published in the *Zeitschrift für Physik,* then picked up the idea and developed it further himself, applying the new statistics not just to particles of light but, under appropriate circumstances, to molecules and atoms.

---

[7] This was before the partition of India and the creation of Pakistan; Dhaka is now the capital of Bangladesh.

The new statistical technique soon became known as "Bose-Einstein statistics," and particles that obey Bose-Einstein statistics are called bosons. Einstein was able to use the new statistics to make predictions about the nature of thermodynamics, and in particular the behavior of certain liquids at very low temperatures, where viscosity disappears and they become "superfluid"; the predicted superfluidity was observed experimentally in 1928, and the behavior of bosons, or "bose condensates," is still a matter of extreme interest to researchers today.

But the most far-reaching aspect of Einstein's work in the mid-1920s came from the analogy between a quantum gas and a molecular gas. Einstein realized that it worked both ways. If the blackbody radiation behaved in the same way as a molecular gas under some circumstances, and if light quanta were also known to have a wavelike nature, the implication was that molecules and other material objects must also have a wavelike nature.

Einstein published this conclusion, derived from Bose-Einstein statistics, in 1925; but there was no need for him to take it further and develop a complete wave theory of matter, because someone else had already come up with the same idea from a different angle. The French physicist Louis de Broglie had finished his PhD thesis in Paris in the spring of 1924, and submitted it to Paul Langevin, an old friend of Einstein. In his thesis, de Broglie suggested that all material particles (electrons and the like) had a wavelength. Langevin was so startled by this suggestion that he asked Einstein to comment on the thesis; Einstein assured Langevin that the thesis was sound, and de Broglie was awarded his doctorate. Five years later, de Broglie received the Nobel Prize for his work.[8] By then, largely thanks to this break-through realization that matter particles also have a wavelike nature, quantum theory was fully established, and Einstein had been in both at the beginning and end of the story. But in 1928 he had suffered another serious illness. In 1929, the year de Broglie received his Nobel Prize, Einstein was fifty years old, no longer a major player in the science game, and beginning to realize that the situation in Germany might soon become untenable.

Although Einstein continued to carry out research and write scientific papers, hardly surprisingly after his fiftieth birthday he did nothing else to rank with his earlier achievements. Most of his later scientific life was devoted to an unsuccessful effort to find a single mathematical package (a unified theory) that would describe both the material world and the world of electromagnetic radiation, echoing the way Maxwell had unified the description of electricity and magnetism into one package. It was a noble effort, but doomed to failure given the limited understanding physicists had of the particle world, in particular, at the time. We shall not describe any of this work but sketch the outlines of Einstein's later years.

Clinging to the hope that the political situation in Germany might yet improve, Einstein avoided resigning from his post in

---

[8] Not least because the wavelengths of electrons had been measured by George Thomson in 1927. George was the son of J. J. Thomson, who discovered the electron.

Berlin as long as he could, but spent as much time as possible out of the country over the next few years. After spending the summer of 1930 in Berlin, he visited Brussels, London, and Zurich (where he received an honorary doctorate from the ETH) in the autumn, returning briefly to Berlin before sailing for California from Antwerp on December 2. The long voyage via New York and the Panama Canal ended in San Diego on December 30, and Einstein stayed as a visiting professor at Caltech, in Pasadena, until mid-March 1931.

Once again, the return to Berlin was brief. In May, Einstein took up an invitation to spend a month giving a series of lectures in Oxford. The visit proved so successful that it led to an invitation for him to become a visiting fellow at one of the Oxford colleges, Christ Church, for five years, with an annual stipend of £400. He could come and go as he pleased, provided he spent some time each year in Oxford. As everyone involved realized, apart from the intrinsic attraction of such an offer, it provided an escape route if things got worse in Germany.

At the same time, Einstein was being courted by Caltech, with an offer of $7,000 for another visit the following winter, and an understanding that the arrangement could be made permanent if he wished. Once again, the Einsteins sailed on December 2, and once again they arrived in California at the end of the year. It was during this visit that Einstein made the contact that would soon result in him finding a settled base.

The contact was with Abraham Flexner, a highly regarded American scientific administrator who had obtained funding for a new research institute in Princeton, and was on a head-hunting expedition to the West Coast looking for eminent scientists to staff it. Who better than the most eminent scientist of the twentieth century? Flexner realized that if he could lure Einstein to his new institute his presence would act as a magnet for other scientists, and ensure the success of the project. But he adopted a cautious approach to snaring his prize.

Einstein returned to Europe in March 1932, committed to spending the next winter at Caltech but with no plans for a permanent move. In the summer, while he was in Oxford, he met up with Flexner (still head-hunting) again. This time, the American offered Einstein a post at what would become the Institute for Advanced Study, on whatever terms Einstein wished. In June, a deal was agreed. Einstein would visit Princeton for six months each year, starting in the autumn of 1933, for a salary of $10,000 and all traveling expenses to be paid by the Institute. The arrangement came just in time. Following elections in July 1932 the Nazis became the largest party in the German government, and the writing was on the wall. Officially, when Albert and Elsa left for their third and last visit to Caltech in December 1932 Einstein was expected back in Berlin to take up his academic duties in April 1933; but, as he locked the door of their house Einstein turned to his wife and said, "Take a very good look at it. You will never see it again."[9]

Adolph Hitler was appointed Reich Chancellor on January

---

[9] Quoted by Philipp Frank, in *Einstein: His Life and Times,* translated by George Rosen, Knopf, New York, 1947.

30, 1933, and in the ensuing wave of anti-Semitism Einstein's bank account was frozen, his house ransacked, and copies of a popular book he had written on relativity were publicly burned. Although the Einsteins returned to Europe at the end of March, they clearly could not return to Germany, but largely divided their time between Oxford and Belgium before leaving for Princeton in October. This was still not seen at the time as a permanent move, and it was expected that Einstein would return to Oxford the following summer. There were even moves to grant him British citizenship. In fact, after 1933 Einstein only left the United States once, traveling to Bermuda in May 1935 in order to make a formal application to reenter the country as a permanent resident.

Soon after taking this step, the Einsteins were able to buy a house in Princeton, at 112 Mercer Street—they had previously been living in a rented apartment. But Elsa did not enjoy the new property for long. She become ill during the summer, and never fully recovered. She died in the house on Mercer Street on December 20, 1936. Einstein, always independently minded, seems to have quickly got over the loss. By this time his divorced stepdaughter, Margot, was acting as his housekeeper and looking after him, while he had an efficient secretary, Helen Dukas, who protected him from outside intrusions. As he wrote to his friend Max Born not long after Elsa died:

I have settled down splendidly here. I hibernate like a bear in its cave, and really feel more at home than ever before in all my varied existence. This bearishness has been accentuated further by the death of my mate, who was more attached to human beings than I.[10]

Over the next few years, most of what was left of Einstein's family also came to America. Hans Albert, who had completed a PhD in engineering at the ETH in 1936, arrived in 1937 with his wife and son, and settled in the United States; he died in 1973. But Eduard, who had developed signs of serious mental illness in the early 1930s, ended his days (in 1965) in a psychiatric hospital in Switzerland. In 1939, Maja and her husband, Paul Winteler, had to leave Italy, where the Fascists were in power. Paul was refused entry to the United States and stayed in Geneva; Maja joined the household in Princeton.

Also in 1939, Einstein played his famous part in alerting the American president, Franklin D. Roosevelt, to the prospect of an atomic bomb. The historic letter to Roosevelt was actually drafted by other scientists, concerned that Hitler's Germany might develop atomic weapons, but Einstein was persuaded to sign the letter and send it to the president, since his name would carry more weight. Very little happened as a result of the letter before the Japanese attack on Pearl Harbor in 1941, but this was the first step in the United States toward the development of the atomic bomb, although Einstein (who became a U.S. citizen on October 1, 1940) played no part in the Manhattan Project itself. This was partly because the authorities involved had doubts about his discretion and were concerned

---

[10] *The Born-Einstein Letters,* Macmillan, London, 1971.

about the left-wing and pacifist sympathies he had expressed earlier in his life.

Einstein's contribution to the American war effort was limited to acting as a consultant for the U.S. Navy, assessing various schemes put forward for new weapons. This was an ideal job for a former technical expert in the Swiss patent office but perhaps did not make full use of his abilities.

After the war, Einstein experienced another bout of serious illness and was never again fully fit. He officially retired in 1945 but kept his office at the Institute and continued to work there whenever he wanted to and felt up to it. Maja had intended to go back to Switzerland when the war ended, but she suffered a stroke and became bedridden; Einstein read to her every day until her death in 1951. By then, Mileva had already died, in Zurich in 1948. It's hardly surprising that when the famous offer of the presidency of Israel came in November 1952, Einstein, now seventy-three, felt unable to accept. In his formal letter turning down the invitation he said, "I lack both the natural aptitude and the experience to deal properly with people and exercise official functions . . . even if advancing age was not making increasing demands on my strength."[11]

In spite of his physical deterioration, however, he remained mentally fit and tried to use the power of his name to nip the nuclear arms race in the bud. After the announcement of the American hydrogen bomb program in 1950, Einstein made a

---

[11] Otto Nathan & Heinz Norden, editors, *Einstein on Peace,* Simon & Schuster, New York, 1960.

televised broadcast in which he warned that unless the continuing development of bigger and "better" bombs were stopped:

> Radioactive poisoning of the atmosphere and, hence, annihilation of all life on Earth will have been brought within the range of what is technically possible. The weird aspect of this development lies in its apparently inexorable character. Each step appears as the inevitable consequence of the one that went before. And at the end, looming ever clearer, lies general annihilation.[12]

Right to the end of his life, Einstein continued to speak out against the nuclear arms race and in defense of the civil liberties attacked in the early 1950s by the McCarthy witch hunts. But that end was not far off. He became ill again in April 1955, not long after his seventy-sixth birthday and a month after the fiftieth anniversary of the completion of the first paper of his annus mirabilis, the paper for which he received the Nobel Prize. Einstein was taken to hospital; but he refused any treatment to prolong his life, describing such intervention as "tasteless."[13] A little after 1 A.M. on April 18, 1955, with only a nurse in attendance, he muttered a few words of German and died. The nurse knew no German.

---

[12] See *Einstein on Peace*.
[13] See Pais.

FURTHER READING

Most of these books are accessible at about the level of the present volume, but go into more detail about Einstein's life or work. Titles marked with an asterisk require a little more scientific background.

Bernstein, Jeremy. *Albert Einstein and the Frontiers of Physics*. Oxford University Press, 1996.

Clark, Ronald. *Einstein: The Life and Times*. London: Hodder & Stoughton, 1973.

*The Collected Papers of Albert Einstein, volumes three and four. Edited by Martin Klein and colleagues. Princeton University Press, 1993 and 1995.

*The Collected Papers of Albert Einstein, volumes five and six. Edited by Martin Klein and colleagues. Princeton University Press, 1993 and 1996.

Einstein, Albert. Relativity. New York: Crown, 1961 (reprint in English of Einstein's only "popular" book; originally published by Holt, New York, 1921).

Einstein, Albert. Autobiographical Notes. Edited and translated by P. A. Schilpp. La Salle, Ill.: Open Court, 1979.

Einstein's Miraculous Year. Edited by John Stachel. Princeton University Press, 1998.

Epstein, Lewis Carroll. Relativity Visualized. Revised edition. San Francisco: Insight Press, 1987.

Fölsing, Albrecht. Albert Einstein. Translated by Ewald Osers. New York: Viking, 1997.

Gamow, George. Mr. Tompkins in Paperback. Cambridge University Press, 1965.

Gribbin, John. In Search of Schrödinger's Cat. New York: Bantam, 1984.

Hey, Tony, and Patrick Walters. Einstein's Mirror. Cambridge University Press, 1997.

Millikan, Robert. The Autobiography, New York: Prentice Hall, 1950. London, 1951.

Overbye, Dennis. Einstein in Love. New York: Viking, 2000.

*Pais, Abraham. Subtle Is the Lord. Oxford University Press, 1982.

Renn, Jürgen, and Robert Schulmann, editors. Albert Einstein, Mileva Maric: The Love Letters. Translated by Shawn Smith. Princeton University Press, 1992.

Seelig, Carl. *Albert Einstein*. Translated by Mervyn Savill. London: Staples Press, 1956.

*Stachel, John, and colleagues (editors). *The Collected Papers of Albert Einstein,* volumes one and two. Princeton University Press, 1987 and 1989.

Stannard, Russell. *The Time and Space of Uncle Albert*. London: Faber, 1989.

White, Michael, and John Gribbin. *Einstein: A Life in Science*. Revised edition. London: Simon & Schuster, 2005.

Other biographies by John & Mary Gribbin:

FitzRoy: The Remarkable Story of Darwin's Captain and the Invention of the Weather Forecast (Yale UP, 2004).

Richard Feynman: A Life in Science (Penguin, London, 1998).

Timeline of Albert Einstein's Life

1879: Albert Einstein is born to Hermann and Pauline Einstein in Ulm, Germany.

1880: Einstein's family moves to Munich, Germany.

1888: Einstein is accepted at the Luitpold-Gymnasium.

1894: The Einstein family moves to Milan, Italy, but Einstein stays in Munich to complete his education. He later drops out and moves to Milan.

1896: Einstein graduates from high school at the age of seventeen and enrolls at the ETH (the Federal Polytechnic) in Zurich. He also renounces his German citizenship.

1898: Einstein falls in love with Mileva Maric, a classmate at the ETH.

1900: Einstein graduates from the ETH as a teacher of mathematics and physics.

1901: Einstein officially becomes a Swiss citizen, works as a temporary teacher at the Technical College in Winterthur, Switzerland, and later works as a teacher at a private school in Schaffhausen, Switzerland. Mileva becomes pregnant.

1902: Lieserl, Einstein and Mileva's illegitimate child, is born in January and later dies or is given up for adoption. In June, Einstein is hired as a patent officer in Bern.

1903: Einstein and Mileva marry.

1904: Hans Albert, Einstein and Mileva's first son, is born.

1905: Annus mirabilis, Einstein's "miracle year," in which he completes a doctoral degree and three papers including the basis of the theory of relativity.

1908: Einstein becomes a lecturer at Bern University.

1910: Einstein and Mileva have a second son, Eduard.

1912: Einstein is appointed professor of theoretical physics at the ETH.

1914: The Einsteins move to Berlin, where Albert takes a position with the University of Berlin. Later, Einstein and Mileva separate, and she returns to Zurich with their sons. Divorce proceedings begin.

1915: Einstein completes the general theory of relativity.

1917: Einstein falls seriously ill and is nursed back to health by his cousin Elsa, whom he later moves in with.

1919: Einstein and Mileva obtain a divorce, and he marries Elsa. A solar eclipse proves Einstein's general theory of relativity, and the news is announced at a meeting of the Royal Society of London.

1922:   Einstein is awarded the Nobel Prize in physics for his work on quantum theory.

1927:   Einstein attends the fifth Solvay Conference in Brussels, and begins debating the foundation of quantum mechanics with Niels Bohr.

1933:   The Nazis seize power in Germany, and Einstein and Elsa move to the United States. They settle in Princeton, New Jersey, where he assumes a full-time post at the Institute of Advanced Study.

1936:   Elsa Einstein dies.

1939:   World War II begins in Europe, and Einstein signs a letter alerting President Franklin D. Roosevelt to the prospect of an atomic bomb.

1940:   Einstein becomes an American citizen.

1949:   Mileva dies.

1955:   Einstein dies on April 18.

Relativity:
The Special and General
Theory by Albert Einstein

CONTENTS

CONTENTS

CONTENTS

(Note: Although originally published in 1916, these papers were
later revised to take into account new developments.)

(December 1916)

The present book is intended, as far as possible, to give an exact insight into the theory of relativity to those readers who, from a general scientific and philosophical point of view, are interested in the theory, but who are not conversant with the mathematical apparatus of theoretical physics. The work presumes a standard of education corresponding to that of a university matriculation examination, and, despite the shortness of the book, a fair amount of patience and force of will on the part of the reader. The author has spared himself no pains in his endeavour to present the main ideas in the simplest and most intelligible form, and on the whole, in the sequence and connection in which they actually originated. In the interest of clearness, it appeared to me inevitable that I should repeat myself frequently, without paying the slightest attention to the elegance of the presentation. I adhered scrupulously to the precept of that brilliant theoretical physicist L. Boltzmann, according to whom matters of elegance ought to be left to the tailor and to the cobbler. I make no pretence of having withheld from the reader difficulties which are inherent to the subject. On the other hand, I have purposely treated the empirical physical foundations of the theory in a "step-motherly" fashion, so that readers unfamiliar with physics may not feel like the wanderer who was unable to see the forest for the trees. May the book bring some one a few happy hours of suggestive thought!

—A. Einstein
December 1916

Abbreviations Used in Text

K = co-ordinate system

x, y = two-dimensional co-ordinates

x, y, z = three-dimensional co-ordinates

x, y, z, t = four-dimensional co-ordinates

t = time

I = distance

v = velocity

F = force

G = gravitational field

PART ONE

The Special Theory of Relativity

ONE

Physical Meaning of
Geometrical Propositions

I n your schooldays most of you who read this book made
acquaintance with the noble building of Euclid's geometry, and you remember—perhaps with more respect than
love—the magnificent structure, on the lofty staircase of which
you were chased about for uncounted hours by conscientious
teachers. By reason of our past experience, you would certainly
regard everyone with disdain who should pronounce even the
most out-of-the-way proposition of this science to be untrue.
But perhaps this feeling of proud certainty would leave you
immediately if some one were to ask you: "What, then, do you

mean by the assertion that these propositions are true?" Let us proceed to give this question a little consideration.

Geometry sets out from certain conceptions such as "plane," "point," and "straight line," with which we are able to associate more or less definite ideas, and from certain simple propositions (axioms) which, in virtue of these ideas, we are inclined to accept as "true." Then, on the basis of a logical process, the justification of which we feel ourselves compelled to admit, all remaining propositions are shown to follow from those axioms, *i.e.* they are proven. A proposition is then correct ("true") when it has been derived in the recognised manner from the axioms. The question of "truth" of the individual geometrical propositions is thus reduced to one of the "truth" of the axioms. Now it has long been known that the last question is not only unanswerable by the methods of geometry, but that it is in itself entirely without meaning. We cannot ask whether it is true that only one straight line goes through two points. We can only say that Euclidean geometry deals with things called "straight lines," to each of which is ascribed the property of being uniquely determined by two points situated on it. The concept "true" does not tally with the assertions of pure geometry, because by the word "true" we are eventually in the habit of designating always the correspondence with a "real" object; geometry, however, is not concerned with the relation of the ideas involved in it to objects of experience, but only with the logical connection of these ideas among themselves.

It is not difficult to understand why, in spite of this, we feel constrained to call the propositions of geometry "true."

Geometrical ideas correspond to more or less exact objects in nature, and these last are undoubtedly the exclusive cause of the genesis of those ideas. Geometry ought to refrain from such a course, in order to give to its structure the largest possible logical unity. The practice, for example, of seeing in a "distance" two marked positions on a practically rigid body is something which is lodged deeply in our habit of thought. We are accustomed further to regard three points as being situated on a straight line, if their apparent positions can be made to coincide for observation with one eye, under suitable choice of our place of observation.

If, in pursuance of our habit of thought, we now supplement the propositions of Euclidean geometry by the single proposition that two points on a practically rigid body always correspond to the same distance (line-interval), independently of any changes in position to which we may subject the body, the propositions of Euclidean geometry then resolve themselves into propositions on the possible relative position of practically rigid bodies.[1] Geometry which has been supplemented in this way is then to be treated as a branch of physics. We can now legitimately ask as to the "truth" of geometrical propositions interpreted in this way, since we are justified in asking whether these propositions are satisfied for those real things we have associated with the geometrical ideas. In less exact terms we can express this by saying that by the "truth" of a geometrical proposition in this sense we understand its validity for a construction with rule and compasses.

Of course the conviction of the "truth" of geometrical

propositions in this sense is founded exclusively on rather incomplete experience. For the present we shall assume the "truth" of the geometrical propositions, then at a later stage (in the general theory of relativity) we shall see that this "truth" is limited, and we shall consider the extent of its limitation.

Note

1. It follows that a natural object is associated also with a straight line. Three points *A*, *B* and *C* on a rigid body thus lie in a straight line when the points *A* and *C* being given, *B* is chosen such that the sum of the distances *AB* and *BC* is as short as possible. This incomplete suggestion will suffice for the present purpose.

TWO

The System of Co-ordinates

O n the basis of the physical interpretation of distance which has been indicated, we are also in a position to establish the distance between two points on a rigid body by means of measurements. For this purpose we require a "distance" (rod S) which is to be used once and for all, and which we employ as a standard measure. If, now, A and B are two points on a rigid body, we can construct the line joining them according to the rules of geometry; then, starting from A, we can mark off the distance S time after time until we reach B. The number of these operations required is the numerical measure of the distance AB. This is the basis of all measurement of length.[1]

Every description of the scene of an event or of the position of an object in space is based on the specification of the point on a rigid body (body of reference) with which that event or object coincides. This applies not only to scientific description, but also to everyday life. If I analyse the place specification "Times Square, New York,"[2] I arrive at the following result. The earth is the rigid body to which the specification of place refers; "Times Square, New York," is a well-defined point, to which a name has been assigned, and with which the event coincides in space.[3]

This primitive method of place specification deals only with places on the surface of rigid bodies, and is dependent on the existence of points on this surface which are distinguishable from each other. But we can free ourselves from both of these limitations without altering the nature of our specification of position. If, for instance, a cloud is hovering over Times Square, then we can determine its position relative to the surface of the earth by erecting a pole perpendicularly on the Square, so that it reaches the cloud. The length of the pole measured with the standard measuring-rod, combined with the specification of the position of the foot of the pole, supplies us with a complete place specification. On the basis of this illustration, we are able to see the manner in which a refinement of the conception of position has been developed.

a. We imagine the rigid body, to which the place specification is referred, supplemented in such a manner that the object whose position we require is reached by the completed rigid body.

b. In locating the position of the object, we make use of a number (here the length of the pole measured with the measuring-rod) instead of designated points of reference.

c. We speak of the height of the cloud even when the pole which reaches the cloud has not been erected. By means of optical observations of the cloud from different positions on the ground, and taking into account the properties of the propagation of light, we determine the length of the pole we should have required in order to reach the cloud.

From this consideration we see that it will be advantageous if, in the description of position, it should be possible by means of numerical measures to make ourselves independent of the existence of marked positions (possessing names) on the rigid body of reference. In the physics of measurement this is attained by the application of the Cartesian system of co-ordinates.

This consists of three plane surfaces perpendicular to each other and rigidly attached to a rigid body. Referred to a system of co-ordinates, the scene of any event will be determined (for the main part) by the specification of the lengths of the three perpendiculars or co-ordinates (x, y, z) which can be dropped from the scene of the event to those three plane surfaces. The lengths of these three perpendiculars can be determined by a series of manipulations with rigid measuring-rods performed according to the rules and methods laid down by Euclidean geometry.

In practice, the rigid surfaces which constitute the system of co-ordinates are generally not available; furthermore, the magnitudes of the co-ordinates are not actually determined by constructions with rigid rods, but by indirect means. If the results of physics and astronomy are to maintain their clearness, the physical meaning of specifications of position must always be sought in accordance with the above considerations.[4]

We thus obtain the following result: Every description of events in space involves the use of a rigid body to which such events have to be referred. The resulting relationship takes for granted that the laws of Euclidean geometry hold for "distances;" the "distance" being represented physically by means of the convention of two marks on a rigid body.

Notes

1. Here we have assumed that there is nothing left over, *i.e.* that the measurement gives a whole number. This difficulty is got over by the use of divided measuring-rods, the introduction of which does not demand any fundamentally new method.

2. Einstein used "Potsdamer Platz, Berlin" in the original text. In the authorised translation this was supplemented with "Trafalgar Square, London." We have changed this to "Times Square, New York," as this is the most well known/identifiable location to English speakers in the present day.

3. It is not necessary here to investigate further the significance of the expression "coincidence in space." This conception is sufficiently obvious to ensure that differences of opinion are scarcely likely to arise as to its applicability in practice.

4. A refinement and modification of these views does not become necessary until we come to deal with the general theory of relativity, treated in the second part of this book.

THREE

Space and Time in Classical Mechanics

The purpose of mechanics is to describe how bodies change their position in space with "time." I should load my conscience with grave sins against the sacred spirit of lucidity were I to formulate the aims of mechanics in this way, without serious reflection and detailed explanations. Let us proceed to disclose these sins.

It is not clear what is to be understood here by "position" and "space." I stand at the window of a railway carriage which is travelling uniformly, and drop a stone on the embankment, without throwing it. Then, disregarding the influence of the air resistance, I see the stone descend in a straight line. A pedestrian who observes the misdeed from the footpath notices that the stone falls to earth in a parabolic curve. I now ask: Do the "positions" traversed by the stone lie "in reality" on a straight line or on a parabola? Moreover, what is meant here by motion "in space"? From the considerations of the previous section the answer is self-evident. In the first place we entirely shun the vague word "space," of which, we must honestly acknowledge, we cannot form the slightest conception, and we replace it by "motion relative to a practically rigid body of reference." The positions relative to the body of reference (railway carriage or embankment) have already been defined in detail in the preceding section. If instead of "body of reference" we insert "system of co-ordinates," which is a useful idea for mathematical description, we are in a position to say: The stone traverses a straight line relative to a system of co-ordinates rigidly attached to the carriage, but relative to a system of co-ordinates rigidly attached to the ground (embankment) it describes a parabola. With the aid of this example it is clearly seen that there is no such thing as an independently existing trajectory (lit. "path-curve"[1]), but only a trajectory relative to a particular body of reference.

In order to have a *complete* description of the motion, we must specify how the body alters its position *with time; i.e.* for every point on the trajectory it must be stated at what time the body is situated there. These data must be supplemented by such a definition of time that, in virtue of this definition, these time-values can be regarded essentially as magnitudes (results of measurements) capable of observation. If we take our stand on the ground of classical mechanics, we can satisfy this requirement for our illustration in the following manner. We imagine two clocks of identical construction; the man at the railway-carriage window is holding one of them, and the man on the footpath the other. Each of the observers determines the position on his own reference-body occupied by the stone at each tick of the clock he is holding in his hand. In this connection we have not taken account of the inaccuracy involved by the finiteness of the velocity of propagation of light. With this and with a second difficulty prevailing here we shall have to deal in detail later.

Note

1. That is, a curve along which the body moves.

FOUR

The Galileian System
of Co-ordinates

As is well known, the fundamental law of the mechanics of Galilei-Newton, which is known as the *law of inertia*, can be stated thus: A body removed sufficiently far from other bodies continues in a state of rest or of uniform motion in a straight line. This law not only says something about the motion of the bodies, but it also indicates the reference-bodies or systems of co-ordinates, permissible in mechanics, which can be used in mechanical description. The visible fixed stars are bodies for which the law of inertia certainly holds to a high degree of approximation. Now if we use a system of co-ordinates which is rigidly attached to the earth, then, relative to this system, every fixed star describes a circle of immense radius in the course of an astronomical day, a result which is opposed to the statement of the law of inertia. So that if we adhere to this law we must refer these motions only to systems of co-ordinates relative to which the fixed stars do not move in a circle. A system of co-ordinates of which the state of motion is such that the law of inertia holds relative to it is called a "Galileian system of co-ordinates." The laws of the mechanics of Galilei-Newton can be regarded as valid only for a Galileian system of co-ordinates.

FIVE

The Principle of Relativity
(in the Restricted Sense)

In order to attain the greatest possible clearness, let us return to our example of the railway carriage supposed to be travelling uniformly. We call its motion a uniform translation ("uniform" because it is of constant velocity and direction, "translation" because although the carriage changes its position relative to the embankment yet it does not rotate in so doing). Let us imagine a raven flying through the air in such a manner that its motion, as observed from the embankment, is uniform and in a straight line. If we were to observe the flying raven from the moving railway carriage, we should find that the motion of the raven would be one of different velocity and direction, but that it would still be uniform and in a straight line. Expressed in an abstract manner we may say: If a mass m is moving uniformly in a straight line with respect to a co-ordinate system K, then it will also be moving uniformly and in a straight line relative to a second co-ordinate system K^1 provided that the latter is executing a uniform translatory motion with respect to K. In accordance with the discussion contained in the preceding section, it follows that:

If K is a Galileian co-ordinate system, then every other co-ordinate system K^1 is a Galileian one, when, in relation to K, it is in a condition of uniform motion of translation. Relative to K^1 the mechanical laws of Galilei-Newton hold good exactly as they do with respect to K.

We advance a step farther in our generalisation when we express the tenet thus: If, relative to K, K^1 is a uniformly moving co-ordinate system devoid of rotation, then natural phenomena run their course with respect to K^1 according to exactly the same general laws as with respect to K. This statement is called the *principle of relativity* (in the restricted sense).

As long as one was convinced that all natural phenomena were capable of representation with the help of classical mechanics, there was no need to doubt the validity of this principle of relativity. But in view of the more recent development of electrodynamics and optics it became more and more evident that classical mechanics affords an insufficient foundation for the physical description of all natural phenomena. At this juncture the question of the validity of the principle of relativity became ripe for discussion, and it did not appear impossible that the answer to this question might be in the negative.

Nevertheless, there are two general facts which at the outset speak very much in favour of the validity of the principle of relativity. Even though classical mechanics does not supply us with a sufficiently broad basis for the theoretical presentation of all physical phenomena, still we must grant it a considerable measure of "truth," since it supplies us with the actual motions of the heavenly bodies with a delicacy of detail little short of wonderful. The principle of relativity must therefore apply with great accuracy in the domain of *mechanics*. But that a principle of such broad generality should hold with such exactness in one domain of phenomena, and yet should be invalid for another, is *a priori* not very probable.

We now proceed to the second argument, to which, moreover, we shall return later. If the principle of relativity (in the restricted sense) does not hold, then the Galileian co-ordinate systems K, K^1, K^2, etc., which are moving uniformly relative to each other, will not be equivalent for the description of natural phenomena. In this case we should be constrained to believe that natural laws are capable of being formulated in a particularly simple manner, and of course only on condition that, from amongst all possible Galileian co-ordinate systems, we should have chosen *one* (K_0) of a particular state of motion as our body of reference. We should then be justified (because of its merits for the description of natural phenomena) in calling this system "absolutely at rest," and all other Galileian systems K "in motion." If, for instance, our embankment were the system K_0

then our railway carriage would be a system K, relative to which less simple laws would hold than with respect to K_0. This diminished simplicity would be due to the fact that the carriage K would be in motion (*i.e.* "really") with respect to K_0. In the general laws of nature which have been formulated with reference to K, the magnitude and direction of the velocity of the carriage would necessarily play a part. We should expect, for instance, that the note emitted by an organ pipe placed with its axis parallel to the direction of travel would be different from that emitted if the axis of the pipe were placed perpendicular to this direction.

Now in virtue of its motion in an orbit round the sun, our earth is comparable with a railway carriage travelling with a velocity of about 30 kilometres per second. If the principle of relativity were not valid we should therefore expect that the direction of motion of the earth at any moment would enter into the laws of nature, and also that physical systems in their behaviour would be dependent on the orientation in space with respect to the earth. For owing to the alteration in direction of the velocity of revolution of the earth in the course of a year, the earth cannot be at rest relative to the hypothetical system K_0 throughout the whole year. However, the most careful observations have never revealed such anisotropic properties in terrestrial physical space, *i.e.* a physical non-equivalence of different directions. This is a very powerful argument in favour of the principle of relativity.

The Theorem of the Addition of Velocities Employed in Classical Mechanics

L et us suppose our old friend the railway carriage to be travelling along the rails with a constant velocity v, and that a man traverses the length of the carriage in the direction of travel with a velocity w. How quickly or, in other words, with what velocity W does the man advance relative to the embankment during the process? The only possible answer seems to result from the following consideration: If the man were to stand still for a second, he would advance relative to the embankment through a distance v equal numerically to the velocity of the carriage. As a consequence of his walking, however, he traverses an additional distance w relative to the carriage, and hence also relative to the embankment, in this second, the distance w being numerically equal to the velocity with which he is walking. Thus in total he covers the distance $W = v + w$ relative to the embankment in the second considered. We shall see later that this result, which expresses the theorem of the addition of velocities employed in classical mechanics, cannot be maintained; in other words, the law that we have just written down does not hold in reality. For the time being, however, we shall assume its correctness.

SEVEN

The Apparent Incompatibility of the Law of Propagation of Light with the Principle of Relativity

There is hardly a simpler law in physics than that according to which light is propagated in empty space. Every child at school knows, or believes he knows, that this propagation takes place in straight lines with a velocity c = 300,000 km./sec. At all events we know with great exactness that this velocity is the same for all colours, because if this were not the case, the minimum of emission would not be observed simultaneously for different colours during the eclipse of a fixed star by its dark neighbour. By means of similar considerations based on observations of double stars, the Dutch astronomer De Sitter was also able to show that the velocity of propagation of light cannot depend on the velocity of motion of the body emitting the light. The assumption that this velocity of propagation is dependent on the direction "in space" is in itself improbable.

In short, let us assume that the simple law of the constancy of the velocity of light c (in vacuum) is justifiably believed by the child at school. Who would imagine that this simple law has plunged the conscientiously thoughtful physicist into the greatest intellectual difficulties? Let us consider how these difficulties arise.

Of course we must refer the process of the propagation of light (and indeed every other process) to a rigid reference-body (co-ordinate system). As such a system let us again choose our embankment. We shall imagine the air above it to have been removed. If a ray of light be sent along the embankment, we see from the above that the tip of the ray will be transmitted with the velocity c relative to the embankment. Now let us suppose that our railway carriage is again travelling along the railway lines with the velocity v, and that its direction is the same as that of the ray of light, but its velocity of course much less. Let us inquire about the velocity of propagation of the ray of light relative to the carriage. It is obvious that we can here apply the consideration of the previous section, since the ray of light plays the part of the man walking along relatively to the carriage. The velocity w of the man relative to the embankment is here replaced by the velocity of light relative to the embankment. w is the required velocity of light with respect to the carriage, and we have

$$w = c - v.$$

The velocity of propagation of a ray of light relative to the carriage thus comes out smaller than c.

But this result comes into conflict with the principle of relativity set forth in Section 5. For, like every other general law of nature, the law of the transmission of light *in vacuo* [in vacuum] must, according to the principle of relativity, be the same for the railway carriage as reference-body as when the rails are the body of reference. But, from our above consideration, this would appear to be impossible. If every ray of light is propagated relative to the embankment with the velocity c, then for this reason it would appear that another law of propagation of light must necessarily hold with respect to the carriage—a result contradictory to the principle of relativity.

In view of this dilemma there appears to be nothing else for it than to abandon either the principle of relativity or the simple law of the propagation of light *in vacuo*. Those of you who have carefully followed the preceding discussion are almost sure to expect that we should retain the principle of relativity, which appeals so convincingly to the intellect because it is so natural and simple. The law of the propagation of light *in vacuo* would then have to be replaced by a more complicated law conformable to the principle of relativity. The development of theoretical physics shows, however, that we cannot pursue this course. The epoch-making theoretical investigations of H. A. Lorentz on the electrodynamical and optical phenomena connected with moving bodies show that experience in this domain leads conclusively to a theory of electromagnetic phenomena, of which the law of the constancy of the velocity of light *in vacuo* is a necessary consequence. Prominent theoretical physicists were therefore more inclined to reject the principle of relativity, in spite of the fact that no empirical data had been found which were contradictory to this principle.

At this juncture the theory of relativity entered the arena. As a result of an analysis of the physical conceptions of time and space, it became evident that *in reality there is not the least incompatibility between the principle of relativity and the law of propagation of light,* and that by systematically holding fast to both these laws a logically rigid theory could be arrived at. This theory has been called the *special theory of relativity* to distinguish it from the extended theory, with which we shall deal later. In the following pages we shall present the fundamental ideas of the special theory of relativity.

EIGHT

On the Idea of Time in Physics

Lightning has struck the rails on our railway embankment at two places A and B far distant from each other. I make the additional assertion that these two lightning flashes occurred simultaneously. If I ask you whether there is sense in this statement, you will answer my question with a decided "Yes." But if I now approach you with the request to explain to me the sense of the statement more precisely, you find after some consideration that the answer to this question is not so easy as it appears at first sight.

After some time perhaps the following answer would occur to you: "The significance of the statement is clear in itself and needs no further explanation; of course it would require some consideration if I were to be commissioned to determine by observations whether in the actual case the two events took place simultaneously or not." I cannot be satisfied with this answer for the following reason. Supposing that as a result of ingenious considerations an able meteorologist were to discover that the lightning must always strike the places A and B simultaneously, then we should be faced with the task of testing whether or not this theoretical result is in accordance with the reality. We encounter the same difficulty with all physical statements in which the conception "simultaneous" plays a part. The concept does not exist for the physicist until he has the possibility of discovering whether or not it is fulfilled in an actual case. We thus require a definition of simultaneity such that this definition supplies us with the method by means of which, in the present case, he can decide by experiment whether or not both the lightning strokes occurred simultaneously. As long as this requirement is not satisfied, I allow myself to be deceived as a physicist (and of course the same applies if I am not a physicist), when I imagine that I am able to attach a meaning to the statement of simultaneity. (I would ask the reader not to proceed farther until he is fully convinced on this point.)

After thinking the matter over for some time you then offer the following suggestion with which to test simultaneity. By measuring along the rails, the connecting line AB should be measured up and an observer placed at the mid-point M of the distance AB. This observer should be supplied with an arrangement (*e.g.* two mirrors inclined at 90°) which allows him visually to observe both places A and B at the same time. If the observer perceives the two flashes of lightning at the same time, then they are simultaneous.

I am very pleased with this suggestion, but for all that I cannot regard the matter as quite settled, because I feel constrained to raise the following objection:

"Your definition would certainly be right, if only I knew that the light by means of which the observer at M perceives the lightning flashes travels along the length A → M with the same velocity as along the length B → M. But an examination of this supposition would only be possible if we already had at our disposal the means of measuring time. It would thus appear as though we were moving here in a logical circle."

After further consideration you cast a somewhat disdainful glance at me—and rightly so—and you declare:

"I maintain my previous definition nevertheless, because in reality it assumes absolutely nothing about light. There is only one demand to be made of the definition of simultaneity, namely, that in every real case it must supply us with an empirical decision as to whether or not the conception that has to be defined is fulfilled. That my definition satisfies this demand is indisputable. That light requires the same time to traverse the path A → M as for the path B → M is in reality neither a *supposition nor a hypothesis* about the physical nature of light, but a stipulation which I can make of my own freewill in order to arrive at a definition of simultaneity."

It is clear that this definition can be used to give an exact meaning not only to *two* events, but to as many events as we care to choose, and independently of the positions of the scenes of the events with respect to the body of reference[1] (here the railway embankment). We are thus led also to a definition of "time" in physics. For this purpose we suppose that clocks of identical construction are placed at the points A, B and C of the railway line (co-ordinate system) and that they are set in such a manner that the positions of their pointers are simultaneously (in the above sense) the same. Under these conditions we understand by the "time" of an event the reading (position of the hands) of that one of these clocks which is in the immediate vicinity (in space) of the event. In this manner a time-value is associated with every event which is essentially capable of observation.

This stipulation contains a further physical hypothesis, the validity of which will hardly be doubted without empirical evidence to the contrary. It has been assumed that all these clocks go *at the same rate* if they are of identical construction. Stated more exactly: When two clocks arranged at rest in different places of a reference-body are set in such a manner that a *particular* position of the pointers of the one clock is *simultaneous* (in the above sense) with the same position, of the pointers of the other clock, then identical "settings" are always simultaneous (in the sense of the above definition).

Note: We suppose further that, when three events *A, B* and *C* take place in different places in such a manner that, if *A* is simultaneous with *B*, and *B* is simultaneous with *C* (simultaneous in the sense of the above definition), then the criterion for the simultaneity of the pair of events *A, C* is also satisfied. This assumption is a physical hypothesis about the law of propagation of light; it must certainly be fulfilled if we are to maintain the law of the constancy of the velocity of light *in vacuo*.

NINE

The Relativity of Simultaneity

U p to now our considerations have been referred to a particular body of reference, which we have styled a "railway embankment." We suppose a very long train travelling along the rails with the constant velocity v and in the direction indicated in Fig 1. People travelling in this train will with a vantage view the train as a rigid reference-body (co-ordinate system); they regard all events in

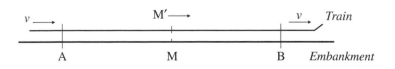

- 178 -

reference to the train. Then every event which takes place along the line also takes place at a particular point of the train. Also the definition of simultaneity can be given relative to the train in exactly the same way as with respect to the embankment. As a natural consequence, however, the following question arises :

Are two events (*e.g.* the two strokes of lightning A and B) which are simultaneous *with reference to the railway embankment* also simultaneous *relatively to the train*? We shall show directly that the answer must be in the negative.

When we say that the lightning strokes A and B are simultaneous with respect to the embankment, we mean: the rays of light emitted at the places A and B, where the lightning occurs, meet each other at the mid-point M of the length A → B of the embankment. But the events A and B also correspond to positions A and B on the train. Let M′ be the mid-point of the distance A → B on the travelling train. Just when the flashes (as judged from the embankment) of lightning occur, this point M′ naturally coincides with the point M but it moves towards the right in the diagram with the velocity v of the train. If an observer sitting in the position M′ in the train did not possess this velocity, then he would remain permanently at M, and the light rays emitted by the flashes of lightning A and B would reach him simultaneously, *i.e.* they would meet just where he is situated. Now in reality (considered with reference to the railway embankment) he is hastening towards the beam of light coming from B, whilst he is riding on ahead of the beam of light coming from A. Hence the observer will see the beam of light emitted from B earlier than he will see that emitted from A. Observers who take the railway train as their reference-body must therefore come to the conclusion that the lightning flash B took place earlier than the lightning flash A. We thus arrive at the important result:

Events which are simultaneous with reference to the embankment are not simultaneous with respect to the train, and *vice versa* (relativity of simultaneity). Every reference-body (co-ordinate system) has its own particular time; unless we are told the reference-body to which the statement of time refers, there is no meaning in a statement of the time of an event.

Now before the advent of the theory of relativity it had always tacitly been assumed in physics that the statement of time had an absolute significance, *i.e.* that it is independent of the state of motion of the body of reference. But we have just seen that this assumption is incompatible with the most natural definition of simultaneity; if we discard this assumption, then the conflict between the law of the propagation of light *in vacuo* and the principle of relativity (developed in Section 7) disappears.

We were led to that conflict by the considerations of Section 6, which are now no longer tenable. In that section we concluded that the man in the carriage, who traverses the distance w *per second* relative to the carriage, traverses the same distance also with respect to the embankment *in each second* of time. But, according to the foregoing considerations, the time required by a particular occurrence with respect to the carriage must not be considered equal to the duration of the same occurrence as judged from the embankment (as reference-body). Hence it cannot be contended that the man in walking travels the distance w relative to the railway line in a time which is equal to one second as judged from the embankment.

Moreover, the considerations of Section 6 are based on yet a second assumption, which, in the light of a strict consideration, appears to be arbitrary, although it was always tacitly made even before the introduction of the theory of relativity.

TEN

On the Relativity of the Conception of Distance

Let us consider two particular points on the train[1] travelling along the embankment with the velocity v, and inquire as to their distance apart. We already know that it is necessary to have a body of reference for the measurement of a distance, with respect to which body the distance can be measured up. It is the simplest plan to use the train itself as reference-body (co-ordinate system). An observer in the train measures the interval by marking off his measuring-rod in a straight line (*e.g.* along the floor of the carriage) as many times as is necessary to take him from the one marked point to the other. Then the number which tells us how often the rod has to be laid down is the required distance.

It is a different matter when the distance has to be judged from the railway line. Here the following method suggests itself. If we call A′ and B′ the two points on the train whose distance apart is required, then both of these points are moving with the velocity v along the embankment. In the first place we require to determine the points A and B of the embankment which are just being passed by the two points A′ and B′ at a particular time t—judged from the embankment. These points A and B of the embankment can be determined by applying the definition of time given in Section 8. The distance between these points A and B is then measured by repeated application of the measuring-rod along the embankment.

A priori it is by no means certain that this last measurement will supply us with the same result as the first. Thus the length of the train as measured from the embankment may be different from that obtained by measuring in the train itself. This circumstance leads us to a second objection which must be raised against the apparently obvious consideration of Section 6. Namely, if the man in the carriage covers the distance w in a unit of time—*measured from the train*—then this distance—*as measured from the embankment*—is not necessarily also equal to w.

Note

1. *E.g.* the middle of the first and of the hundredth carriage.

ELEVEN

The Lorentz Transformation

The results of the last three sections show that the apparent incompatibility of the law of propagation of light with the principle of relativity (Section 7) has been derived by means of a consideration which borrowed two unjustifiable hypotheses from classical mechanics; these are as follows:

1. The time-interval (time) between two events is independent of the condition of motion of the body of reference.
2. The space-interval (distance) between two points of a rigid body is independent of the condition of motion of the body of reference.

If we drop these hypotheses, then the dilemma of Section 7 disappears, because the theorem of the addition of velocities derived in Section 6 becomes invalid. The possibility presents itself that the law of the propagation of light *in vacuo* may be compatible with the principle of relativity, and the question arises: How have we to modify the considerations of Section 6 in order to remove the apparent disagreement between these two fundamental results of experience? This question leads to a general one. In the discussion of Section 6 we have to do with places and times relative both to the train and to the embankment. How are we to find the place and time of an event in relation to the train, when we know the place and time of the event with respect to the railway embankment? Is there a thinkable answer to this question of such a nature that the law of transmission of light *in vacuo* does not contradict the principle of relativity? In other words: Can we conceive of a relation between place and time of the individual events relative to both reference-bodies, such that every ray of light possesses the velocity of transmission c relative to the embankment and relative to the train? This question leads to a quite definite positive answer, and to a perfectly definite transformation law for the space-time magnitudes of an event when changing over from one body of reference to another.

Before we deal with this, we shall introduce the following incidental consideration. Up to the present we have only considered events taking place along the embankment, which had mathematically to assume the function of a straight line. In the manner indicated in Section 2 we can imagine this reference-body

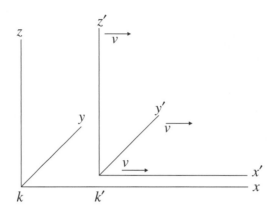

supplemented laterally and in a vertical direction by means of a framework of rods, so that an event which takes place anywhere can be localised with reference to this framework. Similarly, we can imagine the train travelling with the velocity v to be continued across the whole of space, so that every event, no matter how far off it may be, could also be localised with respect to the second framework. Without committing any fundamental error, we can disregard the fact that in reality these frameworks would continually interfere with each other, owing to the impenetrability of solid bodies. In every such framework we imagine three surfaces perpendicular to each other marked out, and designated as "co-ordinate planes" ("co-ordinate system"). A co-ordinate system K then corresponds to the embankment, and a co-ordinate system K[1] to the train. An event, wherever it may have taken place, would be fixed in space with respect to K by the three perpendiculars x, y, z on the co-ordinate planes, and with regard to time by a time value t. Relative to K[1], *the same event* would be fixed in respect of space and time by corresponding values x^1, y^1, z^1, t^1, which of course are not identical with x, y, z, t. It has already been set forth in detail how these magnitudes are to be regarded as results of physical measurements.

Obviously our problem can be exactly formulated in the following manner. What are the values x^1, y^1, z^1, t^1, of an event with respect to K^1, when the magnitudes x, y, z, t of the same event with respect to K are given? The relations must be so chosen that the law of the transmission of light *in vacuo* is satisfied for one and the same ray of light (and of course for every ray) with respect to K and K^1. For the relative orientation in space of the co-ordinate systems indicated in the diagram (Fig. 2), this problem is solved by means of the equations :

$$x' = \frac{x - vt}{\sqrt{I - \dfrac{v^2}{c^2}}}$$

$$y^1 = y$$
$$z^1 = z$$

$$t' - \frac{t - \dfrac{v}{c^2} - x}{\sqrt{I - \dfrac{v^2}{c^2}}}$$

This system of equations is known as the "Lorentz transformation."[1]

If in place of the law of transmission of light we had taken as our basis the tacit assumptions of the older mechanics as to the absolute character of times and lengths, then instead of the above we should have obtained the following equations:

$$x^1 = x - vt$$
$$y^1 = y$$
$$z^1 = z$$
$$t^1 = t$$

This system of equations is often termed the "Galilei transformation." The Galilei transformation can be obtained from the Lorentz transformation by substituting an infinitely large value for the velocity of light c in the latter transformation.

Aided by the following illustration, we can readily see that, in accordance with the Lorentz transformation, the law of the transmission of light *in vacuo* is satisfied both for the reference-body K and for the reference-body K^1. A light-signal is sent along the positive x-axis, and this light-stimulus advances in accordance with the equation

$$x = ct,$$

i.e. with the velocity c. According to the equations of the Lorentz transformation, this simple relation between x and t involves a relation between x^1 and t^1. In point of fact, if we substitute for x the value ct in the first and fourth equations of the Lorentz transformation, we obtain:

$$x' = \frac{(c-v)t}{\sqrt{I - \dfrac{v^2}{c^2}}}$$

$$t' = \frac{\left(I - \dfrac{v}{c}\right)t}{\sqrt{I - \dfrac{v^2}{c^2}}}$$

from which, by division, the expression

$$x^1 = ct^1$$

immediately follows. If referred to the system K^1, the propagation of light takes place according to this equation. We thus see that the velocity of transmission relative to the reference-body K^1 is also equal to c. The same result is obtained for rays of light advancing in any other direction whatsoever. Of course this is not surprising, since the equations of the Lorentz transformation were derived conformably to this point of view.

Note

1. A simple derivation of the Lorentz transformation is given in Appendix I.

TWELVE

The Behaviour of Measuring-Rods and Clocks in Motion

Place a metre-rod in the x¹-axis of K¹ in such a manner that one end (the beginning) coincides with the point $x^1 = 0$ whilst the other end (the end of the rod) coincides with the point $x^1 = I$. What is the length of the metre-rod relatively to the system K? In order to learn this, we need only ask where the beginning of the rod and the end of the rod lie with respect to K at a particular time t of the system K. By means of the first equation of the Lorentz transformation the values of these two points at the time t = 0 can be shown to be

$$x_{(\text{beginning of rod})} = 0 \quad \sqrt{I - \frac{v^2}{c^2}}$$

$$x_{\text{(end of rod)}} = I\sqrt{I - \frac{v^2}{c^2}}$$

the distance between the points being $\sqrt{I - v^2/c^2}$.

But the metre-rod is moving with the velocity v relative to K. It therefore follows that the length of a rigid metre-rod moving in the direction of its length with a velocity v is $\sqrt{I - v^2/c^2}$ of a metre.

The rigid rod is thus shorter when in motion than when at rest, and the more quickly it is moving, the shorter is the rod. For the velocity v = c we should have

$$\sqrt{I - v^2/c^2},$$

and for still greater velocities the square-root becomes imaginary. From this we conclude that in the theory of relativity the velocity c plays the part of a limiting velocity, which can neither be reached nor exceeded by any real body.

Of course this feature of the velocity c as a limiting velocity also clearly follows from the equations of the Lorentz transformation, for these become meaningless if we choose values of v greater than c.

If, on the contrary, we had considered a metre-rod at rest in the x-axis with respect to K, then we should have found that the length of the rod as judged from K^1 would have been $\sqrt{I - v^2/c^2}$; this is quite in accordance with the principle of relativity which forms the basis of our considerations.

A priori it is quite clear that we must be able to learn something about the physical behaviour of measuring-rods and clocks from the equations of transformation, for the magnitudes z, y, x, t, are nothing more nor less than the results of measurements obtainable by means of measuring-rods and clocks. If we had based our considerations on the Galileian transformation we should not have obtained a contraction of the rod as a consequence of its motion.

Let us now consider a seconds-clock which is permanently situated at the origin ($x^1 = 0$) of K^1. $t^1 = 0$ and $t^1 = I$ are two successive ticks of this clock. The first and fourth equations of the Lorentz transformation give for these two ticks :

$$t = 0$$

and

$$t' = \frac{I}{\sqrt{I - \dfrac{v^2}{c^2}}}$$

As judged from K, the clock is moving with the velocity v; as judged from this reference-body, the time which elapses between two strokes of the clock is not one second, but

$$\frac{I}{\sqrt{I - \dfrac{v^2}{c^2}}}$$

seconds, *i.e.* a somewhat larger time. As a consequence of its motion the clock goes more slowly than when at rest. Here also the velocity c plays the part of an unattainable limiting velocity.

THIRTEEN

Theorem of the Addition of Velocities. The Experiment of Fizeau

Now in practice we can move clocks and measuring-rods only with velocities that are small compared with the velocity of light; hence we shall hardly be able to compare the results of the previous section directly with the reality. But, on the other hand, these results must strike you as being very singular, and for that reason I shall now draw another conclusion from the theory, one which can easily be derived from the foregoing considerations, and which has been most elegantly confirmed by experiment.

In Section 6 we derived the theorem of the addition of velocities in one direction in the form which also results from the hypotheses of classical mechanics. This theorem can also be deduced readily from the Galilei transformation (Section 11). In place of the man walking inside the carriage, we introduce a point moving relatively to the co-ordinate system K^1 in accordance with the equation

$$x^1 = wt^1$$

By means of the first and fourth equations of the Galilei transformation we can express x^1 and t^1 in terms of x and t, and we then obtain

$$x = (v + w)t$$

This equation expresses nothing else than the law of motion of the point with reference to the system K (of the man with reference to the embankment). We denote this velocity by the symbol W, and we then obtain, as in Section 6,

$$W = v + w \qquad\qquad (A)$$

But we can carry out this consideration just as well on the basis of the theory of relativity. In the equation

$$x^1 = wt^1 \qquad\qquad (B)$$

we must then express x^1 and t^1 in terms of x and t, making use of the first and fourth equations of the Lorentz transformation. Instead of the equation (A) we then obtain the equation

$$W = \frac{v + w}{1 + \dfrac{vw}{c^2}}$$

which corresponds to the theorem of addition for velocities in one direction according to the theory of relativity. The question now arises as to which of these two theorems is the better in accord with experience. On this point we are enlightened by a most important experiment which the brilliant physicist Fizeau performed more than half a century ago, and which has been repeated since then by some of the best experimental physicists, so that there can be no doubt about its result. The experiment is concerned with the following question. Light travels in a motionless liquid with a particular velocity w. How quickly does it travel in the direction of the arrow in the tube T (see the accompanying diagram, Fig. 3) when the liquid above mentioned is flowing through the tube with a velocity v?

In accordance with the principle of relativity we shall certainly have to take for granted that the propagation of light always takes place with the same velocity w *with respect to the liquid,* whether the latter is in motion with reference to other bodies or not. The velocity of light relative to the liquid and the velocity of the latter relative to the tube are thus known, and we require the velocity of light relative to the tube.

It is clear that we have the problem of Section 6 again before us. The tube plays the part of the railway embankment or of the co-ordinate system K, the liquid plays the part of the carriage or of the co-ordinate system K^1, and finally, the light plays the part of the

man walking along the carriage, or of the moving point in the present section. If we denote the velocity of the light relative to the tube by W, then this is given by the equation (A) or (B), according as the Galilei transformation or the Lorentz transformation corresponds to the facts. Experiment[1] decides in favour of equation (B) derived from the theory of relativity, and the agreement is, indeed, very exact. According to recent and most excellent measurements by Zeeman, the influence of the velocity of flow v on the propagation of light is represented by formula (B) to within one percent.

Nevertheless we must now draw attention to the fact that a theory of this phenomenon was given by H. A. Lorentz long before the statement of the theory of relativity. This theory was of a purely electrodynamical nature, and was obtained by the use of particular hypotheses as to the electromagnetic structure of matter. This circumstance, however, does not in the least diminish the conclusiveness of the experiment as a crucial test in favour of the theory of relativity, for the electrodynamics of Maxwell-Lorentz, on which the original theory was based, in no way opposes the theory of relativity. Rather has the latter been developed from electrodynamics as an astoundingly simple combination and generalisation of the hypotheses, formerly independent of each other, on which electrodynamics was built.

Note

1. Fizeau found $W = w + v\left(I - \dfrac{I}{n^2}\right)$, where $n = \dfrac{c}{w}$ is the index of refraction of the liquid. On the other hand, owing to the smallness of $\dfrac{vw}{c^2}$ as compared with I, we can replace (B) in the first place by $W = (w + v)\left(I - \dfrac{vw}{c^2}\right)$, or to the same order of approximation by $w + v\left(I - \dfrac{I}{n^2}\right)$, which agrees with Fizeau's result.

FOURTEEN

The Heuristic Value of the Theory of Relativity

O ur train of thought in the foregoing pages can be epitomised in the following manner. Experience has led to the conviction that, on the one hand, the principle of relativity holds true and that on the other hand the velocity of transmission of light *in vacuo* has to be considered equal to a constant c. By uniting these two postulates we obtained the law of transformation for the rectangular coordinates x, y, z and the time t of the events which constitute the processes of nature. In this connection we did not obtain the Galilei transformation, but, differing from classical mechanics, the *Lorentz transformation*.

The law of transmission of light, the acceptance of which is justified by our actual knowledge, played an important part in this process of thought. Once in possession of the Lorentz transformation, however, we can combine this with the principle of relativity, and sum up the theory thus:

Every general law of nature must be so constituted that it is transformed into a law of exactly the same form when, instead of the space-time variables x, y, z, t of the original co-ordinate system K, we introduce new space-time variables x^1, y^1, z^1, t^1 of a co-ordinate system K^1. In this connection the relation between the ordinary and the accented magnitudes is given by the Lorentz transformation. Or in brief: General laws of nature are co-variant with respect to Lorentz transformations.

This is a definite mathematical condition that the theory of relativity demands of a natural law, and in virtue of this, the theory becomes a valuable heuristic aid in the search for general laws of nature. If a general law of nature were to be found which did not satisfy this condition, then at least one of the two fundamental assumptions of the theory would have been disproved. Let us now examine what general results the latter theory has hitherto evinced.

FIFTEEN

General Results of the Theory

I t is clear from our previous considerations that the (special) theory of relativity has grown out of electrodynamics and optics. In these fields it has not appreciably altered the predictions of theory, but it has considerably simplified the theoretical structure, *i.e.* the derivation of laws, and—what is incomparably more important—it has considerably reduced the number of independent hypotheses forming the basis of theory. The special theory of relativity has rendered the Maxwell-Lorentz theory so plausible, that the latter would have been generally accepted by physicists even if experiment had decided less unequivocally in its favour.

Classical mechanics required to be modified before it could come into line with the demands of the special theory of relativity. For the main part, however, this modification affects only the laws for rapid motions, in which the velocities of matter v are not very small as compared with the velocity of light. We have experience of such rapid motions only in the case of electrons and ions; for other motions the variations from the laws of classical mechanics are too small to make themselves evident in practice. We shall not consider the motion of stars until we come to speak of the general theory of relativity. In accordance with the theory of relativity the kinetic energy of a material point of mass m is no longer given by the well-known expression

$$m \frac{v^2}{2}$$

but by the expression

$$\frac{mc^2}{\sqrt{I - \dfrac{v^2}{c^2}}}$$

This expression approaches infinity as the velocity v approaches the velocity of light c. The velocity must therefore always remain less than c, however great may be the energies used to produce the acceleration. If we develop the expression for the kinetic energy in the form of a series, we obtain

$$mc^2 + m\,\frac{v^2}{2} + \frac{3}{8}\,m\,\frac{v^4}{c^2} + \cdots .$$

When $\frac{v^2}{c^2}$ is small compared with unity, the third of these terms is always small in comparison with the second, which last is alone considered in classical mechanics. The first term mc^2 does not contain the velocity, and requires no consideration if we are only dealing with the question as to how the energy of a point-mass depends on the velocity. We shall speak of its essential significance later.

The most important result of a general character to which the special theory of relativity has led is concerned with the conception of mass. Before the advent of relativity, physics recognised two conservation laws of fundamental importance, namely, the law of the conservation of energy and the law of the conservation of mass; these two fundamental laws appeared to be quite independent of each other. By means of the theory of relativity they have been united into one law. We shall now briefly consider how this unification came about, and what meaning is to be attached to it.

The principle of relativity requires that the law of the conservation of energy should hold not only with reference to a co-ordinate system K, but also with respect to every co-ordinate system K^1 which is in a state of uniform motion of translation relative to K, or, briefly, relative to every "Galileian" system of co-ordinates. In contrast to classical mechanics, the Lorentz transformation is the deciding factor in the transition from one such system to another.

By means of comparatively simple considerations we are led to draw the following conclusion from these premises, in conjunction with the fundamental equations of the electrodynamics of Maxwell: A body moving with the velocity v, which absorbs[1] an amount of energy E_0 in the form of radiation without suffering an alteration in velocity in the process, has, as a consequence, its energy increased by an amount

$$\frac{E_0}{\sqrt{I - \dfrac{v^2}{c^2}}}$$

In consideration of the expression given above for the kinetic energy of the body, the required energy of the body comes out to be

$$\frac{\left(m + \dfrac{E_0}{c^2}\right)c^2}{\sqrt{I - \dfrac{v^2}{c^2}}}$$

Thus the body has the same energy as a body of mass

$$m + \frac{E_0}{c^2}$$

moving with the velocity v. Hence we can say: If a body takes up an amount of energy E_0, then its inertial mass increases by an amount

$$\frac{E_0}{c^2}$$

the inertial mass of a body is not a constant but varies according to the change in the energy of the body. The inertial mass of a system of bodies can even be regarded as a measure of its energy. The law of the conservation of the mass of a system becomes identical with the law of the conservation of energy, and is only valid provided that the system neither takes up nor sends out energy. Writing the expression for the energy in the form

$$\frac{mc^2 + E_0}{\sqrt{I - \dfrac{v^2}{c^2}}}$$

we see that the term mc^2, which has hitherto attracted our attention, is nothing else than the energy possessed by the body[2] before it absorbed the energy E_0.

A direct comparison of this relation with experiment is not possible at the present time (1920; see note[3]), owing to the fact that the changes in energy E_0 to which we can subject a system are not large enough to make themselves perceptible as a change in the inertial mass of the system.

$$\frac{E_0}{c^2}$$

is too small in comparison with the mass m, which was present before the alteration of the energy. It is owing to this circumstance that classical mechanics was able to establish successfully the conservation of mass as a law of independent validity.

Let me add a final remark of a fundamental nature. The success of the Faraday-Maxwell interpretation of electromagnetic action at a distance resulted in physicists becoming convinced that there are no such things as instantaneous actions at a distance (not involving an intermediary medium) of the type of Newton's law of gravitation. According to the theory of relativity, action at a distance with the velocity of light always takes the place of instantaneous action at a distance or of action at a distance with an infinite velocity of transmission. This is connected with the fact that the velocity c plays a fundamental role in this theory. In Part II we shall see in what way this result becomes modified in the general theory of relativity.

Notes

1. E_0 is the energy taken up, as judged from a co-ordinate system moving with the body.

2. As judged from a co-ordinate system moving with the body.

3. The equation $E = mc^2$ has been thoroughly proved time and again since this time.

SIXTEEN

Experience and the Special Theory of Relativity

To what extent is the special theory of relativity supported by experience? This question is not easily answered for the reason already mentioned in connection with the fundamental experiment of Fizeau. The special theory of relativity has crystallised out from the Maxwell-Lorentz theory of electromagnetic phenomena. Thus all facts of experience which support the electromagnetic theory also support the theory of relativity. As being of particular importance, I mention here the fact that the theory of relativity enables us to predict the effects produced on the light reaching us from the fixed stars. These results are obtained in an exceedingly simple manner, and the effects indicated, which are due to the relative motion of the earth with reference to those fixed stars are found to be in accord with experience. We refer to the yearly movement of the apparent position of the fixed stars resulting from the motion of the earth round the sun (aberration), and to the influence of the radial components of the relative motions of the fixed stars with respect to the earth on the colour of the light reaching us from them. The latter effect manifests itself in a slight displacement of the spectral lines of the light transmitted to us from a fixed star, as compared with the position of the same spectral lines when they are produced by a terrestrial source of light (Doppler principle). The experimental arguments in favour of the Maxwell-Lorentz theory, which are at the same time arguments in favour of the theory of relativity, are too numerous to be set forth here. In reality they limit the theoretical possibilities to such an extent, that no other theory than that of Maxwell and Lorentz has been able to hold its own when tested by experience.

But there are two classes of experimental facts hitherto obtained which can be represented in the Maxwell-Lorentz theory only by the introduction of an auxiliary hypothesis, which in itself—*i.e.* without making use of the theory of relativity— appears extraneous.

It is known that cathode rays and the so-called β-rays emitted by radioactive substances consist of negatively electrified particles (electrons) of very small inertia and large velocity. By examining the deflection of these rays under the influence of electric and magnetic fields, we can study the law of motion of these particles very exactly.

In the theoretical treatment of these electrons, we are faced with the difficulty that electrodynamic theory of itself is unable to give an account of their nature. For since electrical masses of one sign repel each other, the negative electrical masses constituting the electron would necessarily be scattered under the influence of their mutual repulsions, unless there are forces of another kind operating between them, the nature of which has hitherto remained obscure to us.[1] If we now assume that the relative distances between the electrical masses constituting the electron remain unchanged during the motion of the electron (rigid connection in the sense of classical mechanics), we arrive at a law of motion of the electron which does not agree with experience. Guided by purely formal points of view, H. A. Lorentz was the first to introduce the hypothesis that the form of the electron experiences a contraction in the direction of motion in consequence of that motion, the contracted length being proportional to the expression

$$\sqrt{I - \frac{v^2}{c^2}}$$

This, hypothesis, which is not justifiable by any electrodynamical facts, supplies us then with that particular law of motion which has been confirmed with great precision in recent years.

The theory of relativity leads to the same law of motion, without requiring any special hypothesis whatsoever as to the structure and the behaviour of the electron. We arrived at a similar conclusion in Section 13 in connection with the experiment of Fizeau, the result of which is foretold by the theory of relativity without the necessity of drawing on hypotheses as to the physical nature of the liquid.

The second class of facts to which we have alluded has reference to the question whether or not the motion of the earth in space can be made perceptible in terrestrial experiments. We have already remarked in Section 5 that all attempts of this nature led to a negative result. Before the theory of relativity was put forward, it was difficult to become reconciled to this negative result, for reasons now to be discussed. The inherited prejudices about time and space did not allow any doubt to arise as to the prime importance of the Galileian transformation for changing over from one body of reference to another. Now assuming that the Maxwell-Lorentz equations hold for a reference-body K, we then find that they do not hold for a reference-body K^1 moving uniformly with respect to K, if we assume that the relations of the Galileian transformation exist between the co-ordinates of K and K^1. It thus appears that, of all Galileian co-ordinate systems, one (K) corresponding to a particular state of motion is physically unique. This result was interpreted physically by regarding K as at rest with respect to a hypothetical æther of space. On the other hand, all coordinate systems K^1 moving relatively to K were to be regarded as in motion with respect to the æther. To this motion of K^1 against the æther ("æther-drift" relative to K^1) were attributed the more complicated laws which were supposed to hold relative to K^1. Strictly speaking, such an æther-drift ought also to be assumed relative to the earth, and for a long time the efforts of physicists were devoted to attempts to detect the existence of an æther-drift at the earth's surface.

In one of the most notable of these attempts Michelson devised a method which appears as though it must be decisive. Imagine two mirrors so arranged on a rigid body that the reflecting surfaces face each other. A ray of light requires a perfectly definite time T to pass from one mirror to the other and back again, if the whole system be at rest with respect to the æther. It is found by calculation, however, that a slightly different time T^1 is required for this process, if the body, together with the mirrors, be moving relatively to the æther. And yet another point: it is shown by calculation that for a given velocity v with reference to the æther, this time T^1 is different when the body is moving perpendicularly to the planes of the mirrors from that resulting when the motion is parallel to these planes. Although the estimated difference between these two times is exceedingly small, Michelson and Morley performed an experiment involving interference in which this difference should have been clearly detectable. But the experiment gave a negative result—a fact very perplexing to physicists. Lorentz and FitzGerald rescued the theory from this difficulty by assuming that the motion of the body relative to the æther produces a contraction of the body in the direction of motion, the amount of contraction being just sufficient to compensate for the difference in time mentioned above. Comparison with the discussion in Section 11 shows that also from the standpoint of the theory of relativity this solution of the difficulty was the right one. But on the basis of the theory of relativity the method of interpretation is incomparably more satisfactory. According to this theory there is no such thing as a "specially favoured" (unique) co-ordinate system to occasion the introduction of the æther-idea, and hence there can be no æther-drift, nor any experiment with which to demonstrate it. Here the contraction of moving bodies follows from the two fundamental principles of the theory, without the introduction of particular hypotheses; and as the prime factor involved in this contraction we find, not the motion in itself, to which we cannot attach any meaning, but the motion with respect to the body of reference chosen in the particular case in point. Thus for a co-ordinate system moving with the earth the mirror system of Michelson and Morley is not shortened, but it is shortened for a co-ordinate system which is at rest relatively to the sun.

Note

1. The general theory of relativity renders it likely that the electrical masses of an electron are held together by gravitational forces.

SEVENTEEN

Minkowski's
Four-Dimensional Space

T he non-mathematician is seized by a mysterious shud-
dering when he hears of "four-dimensional" things, by
a feeling not unlike that awakened by thoughts of the
occult. And yet there is no more commonplace statement than
that the world in which we live is a four-dimensional space-
time continuum.

Space is a three-dimensional continuum. By this we mean
that it is possible to describe the position of a point (at rest) by
means of three numbers (co-ordinates) x, y, z, and that there is
an indefinite number of points in the neighbourhood of this

one, the position of which can be described by co-ordinates such as x_1, y_1, z_1, which may be as near as we choose to the respective values of the co-ordinates x, y, z, of the first point. In virtue of the latter property we speak of a "continuum," and owing to the fact that there are three co-ordinates we speak of it as being "three-dimensional."

Similarly, the world of physical phenomena which was briefly called "world" by Minkowski is naturally four-dimensional in the space-time sense. For it is composed of individual events, each of which is described by four numbers, namely, three space co-ordinates x, y, z, and a time co-ordinate, the time value t. The "world" is in this sense also a continuum; for to every event there are as many "neighbouring" events (realised or at least thinkable) as we care to choose, the co-ordinates x_1, y_1, z_1, t_1, of which differ by an indefinitely small amount from those of the event x, y, z, t originally considered. That we have not been accustomed to regard the world in this sense as a four-dimensional continuum is due to the fact that in physics, before the advent of the theory of relativity, time played a different and more independent role, as compared with the space co-ordinates. It is for this reason that we have been in the habit of treating time as an independent continuum. As a matter of fact, according to classical mechanics, time is absolute, *i.e.* it is independent of the position and the condition of motion of the system of co-ordinates. We see this expressed in the last equation of the Galileian transformation ($t^1 = t$).

The four-dimensional mode of consideration of the "world"

is natural on the theory of relativity, since according to this theory time is robbed of its independence. This is shown by the fourth equation of the Lorentz transformation:

$$t' = \frac{t - \frac{v}{c^2}x}{\sqrt{I - \frac{v^2}{c^2}}}$$

Moreover, according to this equation the time difference Δt^1 of two events with respect to K^1 does not in general vanish, even when the time difference Δt^1 of the same events with reference to K vanishes. Pure "space-distance" of two events with respect to K results in "time-distance" of the same events with respect to K. But the discovery of Minkowski, which was of importance for the formal development of the theory of relativity, does not lie here. It is to be found rather in the fact of his recognition that the four-dimensional space-time continuum of the theory of relativity, in its most essential formal properties, shows a pronounced relationship to the three-dimensional continuum of Euclidean geometrical space.[1] In order to give due prominence to this relationship, however, we must replace the usual time co-ordinate t by an imaginary magnitude $\sqrt{-I} \cdot ct$ proportional to it. Under these conditions, the natural laws satisfying the demands of the (special) theory of relativity assume mathematical forms, in which the time co-ordinate plays exactly the same role as the three space co-ordinates. Formally, these four co-ordinates correspond exactly to the three space co-ordinates in Euclidean geometry. It must be clear even to the non-mathematician that, as a consequence of this purely formal addition to our knowledge, the theory perforce gained clearness in no mean measure.

These inadequate remarks can give the reader only a vague notion of the important idea contributed by Minkowski. Without it the general theory of relativity, of which the fundamental ideas are developed in the following pages, would perhaps have got no farther than its long clothes. Minkowski's work is doubtless difficult of access to anyone inexperienced in mathematics, but since it is not necessary to have a very exact grasp of this work in order to understand the fundamental ideas of either the special or the general theory of relativity, I shall leave it here at present, and revert to it only towards the end of Part II.

Note

1. Cf. the somewhat more detailed discussion in Appendix II.

PART TWO

The General Theory
of Relativity

EIGHTEEN

Special and General Principle of Relativity

The basal principle, which was the pivot of all our previous considerations, was the *special* principle of relativity, *i.e.* the principle of the physical relativity of all *uniform* motion. Let as once more analyse its meaning carefully.

It was at all times clear that, from the point of view of the idea it conveys to us, every motion must be considered only as a relative motion. Returning to the illustration we have frequently used of the embankment and the railway carriage, we can express the fact of the motion here taking place in the following two forms, both of which are equally justifiable :

a. The carriage is in motion relative to the embankment,

b. The embankment is in motion relative to the carriage.

In (a) the embankment, in (b) the carriage, serves as the body of reference in our statement of the motion taking place. If it is simply a question of detecting or of describing the motion involved, it is in principle immaterial to what reference-body we refer the motion. As already mentioned, this is self-evident, but it must not be confused with the much more comprehensive statement called "the principle of relativity," which we have taken as the basis of our investigations.

The principle we have made use of not only maintains that we may equally well choose the carriage or the embankment as our reference-body for the description of any event (for this, too, is self-evident). Our principle rather asserts what follows: If we formulate the general laws of nature as they are obtained from experience, by making use of

a. the embankment as reference-body,
b. the railway carriage as reference-body,

then these general laws of nature (*e.g.* the laws of mechanics or the law of the propagation of light *in vacuo*) have exactly the same form in both cases. This can also be expressed as follows: For the physical description of natural processes, neither of the reference bodies K, K^1 is unique (*lit.* "specially marked out") as compared with the other. Unlike the first, this latter statement need not of necessity hold *a priori*; it is not contained in the conceptions of "motion" and "reference-body" and derivable from them; only *experience* can decide as to its correctness or incorrectness.

Up to the present, however, we have by no means maintained

the equivalence of *all* bodies of reference K in connection with the formulation of natural laws. Our course was more on the following lines. In the first place, we started out from the assumption that there exists a reference-body K, whose condition of motion is such that the Galileian law holds with respect to it: A particle left to itself and sufficiently far removed from all other particles moves uniformly in a straight line. With reference to K (Galileian reference-body) the laws of nature were to be as simple as possible. But in addition to K, all bodies of reference K^1 should be given preference in this sense, and they should be exactly equivalent to K for the formulation of natural laws, provided that they are in a state of *uniform rectilinear and non-rotary motion* with respect to K; all these bodies of reference are to be regarded as Galileian reference-bodies. The validity of the principle of relativity was assumed only for these reference-bodies, but not for others (*e.g.* those possessing motion of a different kind). In this sense we speak of the special principle of relativity, or special theory of relativity.

In contrast to this we wish to understand by the "general principle of relativity" the following statement: All bodies of reference K, K^1, etc., are equivalent for the description of natural phenomena (formulation of the general laws of nature), whatever may be their state of motion. But before proceeding farther, it ought to be pointed out that this formulation must be replaced later by a more abstract one, for reasons which will become evident at a later stage.

Since the introduction of the special principle of relativity has been justified, every intellect which strives after generalisation must feel the temptation to venture the step towards the general principle of relativity. But a simple and apparently quite reliable consideration seems to suggest that, for the present at any rate, there is little hope of success in such an attempt. Let us imagine ourselves transferred to our old friend the railway carriage, which is travelling at a uniform rate. As long as it is moving uniformly, the occupant of the carriage is not sensible of its motion, and it is for this reason that he can without reluctance interpret the facts of the case as indicating that the carriage is at rest, but the embankment in motion. Moreover, according to the special principle of relativity, this interpretation is quite justified also from a physical point of view.

If the motion of the carriage is now changed into a non-uniform motion, as for instance by a powerful application of the brakes, then the occupant of the carriage experiences a correspondingly powerful jerk forwards. The retarded motion is manifested in the mechanical behaviour of bodies relative to the person in the railway carriage. The mechanical behaviour is different from that of the case previously considered, and for this reason it would appear to be impossible that the same mechanical laws hold relatively to the non-uniformly moving carriage, as hold with reference to the carriage when at rest or in uniform motion. At all events it is clear that the Galileian law does not hold with respect to the non-uniformly moving carriage. Because of this, we feel compelled at the present juncture to grant a kind of absolute physical reality to non-uniform motion, in opposition to the general principle of relativity. But in what follows we shall soon see that this conclusion cannot be maintained.

NINETEEN

The Gravitational Field

I f we pick up a stone and then let it go, why does it fall to the ground?" The usual answer to this question is: "Because it is attracted by the earth." Modern physics formulates the answer rather differently for the following reason. As a result of the more careful study of electromagnetic phenomena, we have come to regard action at a distance as a process impossible without the intervention of some intermediary medium. If, for instance, a magnet attracts a piece of iron, we cannot be content to regard this as meaning that the magnet acts directly on the iron through the intermediate empty space, but we are constrained to imagine—after the manner of Faraday—that the magnet always calls into being something physically real in the space around it, that something being what we call a "magnetic field." In its turn this magnetic field operates on the piece of iron, so that the latter strives to move towards the magnet. We shall not discuss here the justification for this incidental conception, which is indeed a somewhat arbitrary one. We shall only mention that with its aid electromagnetic phenomena can be theoretically represented much more satisfactorily than without it, and this applies particularly to the transmission of electromagnetic waves. The effects of gravitation also are regarded in an analogous manner.

The action of the earth on the stone takes place indirectly. The earth produces in its surrounding a gravitational field, which acts on the stone and produces its motion of fall. As we know from experience, the intensity of the action on a body diminishes according to a quite definite law, as we proceed farther and farther away from the earth. From our point of view this means: The law governing the properties of the gravitational field in space must be a perfectly definite one, in order correctly to represent the diminution of gravitational action with the distance from operative bodies. It is something like this: The body (*e.g.* the earth) produces a field in its immediate neighbourhood directly; the intensity and direction of the field at points farther removed from the body are thence determined by the law which governs the properties in space of the gravitational fields themselves.

In contrast to electric and magnetic fields, the gravitational field exhibits a most remarkable property, which is of fundamental importance for what follows. Bodies which are moving under the sole influence of a gravitational field receive an acceleration, *which does not in the least depend either on the material or on the physical state of the body.* For instance, a piece of lead and a piece of wood fall in exactly the same manner in a gravitational field (*in vacuo*), when they start off from rest or with the same initial velocity. This law, which holds most accurately, can be expressed in a different form in the light of the following consideration.

According to Newton's law of motion, we have

$$(\text{Force}) = (\text{inertial mass}) \times (\text{acceleration}),$$

where the "inertial mass" is a characteristic constant of the accelerated body. If now gravitation is the cause of the acceleration, we then have

$$(\text{Force}) = (\text{gravitational mass}) \times (\text{intensity of the gravitational field}),$$

where the "gravitational mass" is likewise a characteristic constant for the body. From these two relations follows:

$$(\text{acceleration}) = \frac{(\text{gravitational mass})}{(\text{inertial mass})} \times (\text{intensity of the gravitational field}).$$

If now, as we find from experience, the acceleration is to be independent of the nature and the condition of the body and always the same for a given gravitational field, then the ratio of the gravitational to the inertial mass must likewise be the same for all bodies. By a suitable choice of units we can thus make this ratio equal to unity. We then have the following law: The *gravitational* mass of a body is equal to its *inertial* law.

It is true that this important law had hitherto been recorded in mechanics, but it had not been *interpreted*. A satisfactory interpretation can be obtained only if we recognise the following fact: *The same* quality of a body manifests itself according to circumstances as "inertia" or as "weight" (*lit.* "heaviness"). In the following section we shall show to what extent this is actually the case, and how this question is connected with the general postulate of relativity.

TWENTY

The Equality of Inertial and Gravitational Mass as an Argument for the General Postulate of Relativity

We imagine a large portion of empty space, so far removed from stars and other appreciable masses, that we have before us approximately the conditions required by the fundamental law of Galilei. It is then possible to choose a Galileian reference-body for this part of space (world), relative to which points at rest remain at rest and points in motion continue permanently in uniform rectilinear motion. As reference-body let us imagine a spacious chest resembling a room with an observer inside who is equipped with apparatus. Gravitation naturally does not exist for this observer. He must fasten himself with strings to the floor, otherwise the slightest impact against the floor will cause him to rise slowly towards the ceiling of the room.

To the middle of the lid of the chest is fixed externally a hook with rope attached, and now a "being" (what kind of a being is immaterial to us) begins pulling at this with a constant force. The chest together with the observer then begin to move "upwards" with a uniformly accelerated motion. In course of time their velocity will reach unheard-of values—provided that we are viewing all this from another reference-body which is not being pulled with a rope.

But how does the man in the chest regard the process? The acceleration of the chest will be transmitted to him by the reaction of the floor of the chest. He must therefore take up this pressure by means of his legs if he does not wish to be laid out full length on the floor. He is then standing in the chest in exactly the same way as anyone stands in a room of a home on our earth. If he releases a body which he previously had in his land, the acceleration of the chest will no longer be transmitted to this body, and for this reason the body will approach the floor of the chest with an accelerated relative motion. The observer will further convince himself *that the acceleration of the body towards the floor of the chest is always of the same magnitude, whatever kind of body he may happen to use for the experiment.*

Relying on his knowledge of the gravitational field (as it was discussed in the preceding section), the man in the chest will thus come to the conclusion that he and the chest are in a gravitational field which is constant with regard to time. Of course he will be puzzled for a moment as to why the chest does not fall in this gravitational field. Just then, however, he discovers the hook in the middle of the lid of the chest and the rope which is attached to it, and he consequently comes to the conclusion that the chest is suspended at rest in the gravitational field.

Ought we to smile at the man and say that he errs in his conclusion? I do not believe we ought to if we wish to remain consistent; we must rather admit that his mode of grasping the situation violates neither reason nor known mechanical laws. Even though it is being accelerated with respect to the "Galileian space" first considered, we can nevertheless regard the chest as being at rest. We have thus good grounds for extending the principle of relativity to include bodies of reference which are accelerated with respect to each other, and as a result we have gained a powerful argument for a generalised postulate of relativity.

We must note carefully that the possibility of this mode of interpretation rests on the fundamental property of the gravitational field of giving all bodies the same acceleration, or, what comes to the same thing, on the law of the equality of inertial and gravitational mass. If this natural law did not exist, the man in the accelerated chest would not be able to interpret the behaviour of the bodies around him on the supposition of a gravitational field, and he would not be justified on the grounds of experience in supposing his reference-body to be "at rest."

Suppose that the man in the chest fixes a rope to the inner side of the lid, and that he attaches a body to the free end of the rope. The result of this will be to strech the rope so that it will hang "vertically" downwards. If we ask for an opinion of the cause of tension in the rope, the man in the chest will say: "The suspended body experiences a downward force in the gravitational field, and this is neutralised by the tension of the rope; what determines the magnitude of the tension of the rope is the *gravitational mass* of the suspended body." On the other hand, an observer who is poised freely in space will interpret the condition of things thus: "The rope must perforce take part in the accelerated motion of the chest, and it transmits this motion to the body attached to it. The tension of the rope is just large enough to effect the acceleration of the body. That which determines the magnitude of the tension of the rope is the *inertial mass* of the body." Guided by this example, we see that our extension of the principle of relativity implies the *necessity* of the law of the equality of inertial and gravitational mass. Thus we have obtained a physical interpretation of this law.

From our consideration of the accelerated chest we see that a general theory of relativity must yield important results on the laws of gravitation. In point of fact, the systematic pursuit of the general idea of relativity has supplied the laws satisfied by the gravitational field. Before proceeding farther, however, I must warn the reader against a misconception suggested by these considerations. A gravitational field exists for the man in the chest, despite the fact that there was no such field for the co-ordinate system first chosen. Now we might easily suppose that the existence of a gravitational field is always only an *apparent* one. We might also think that, regardless of the kind of gravitational field which may be present, we could always choose another reference-body such that *no* gravitational field exists with reference to it. This is by no means true for all gravitational fields, but only for those of quite special form. It is, for instance, impossible to choose a body of reference such that, as judged from it, the gravitational field of the earth (in its entirety) vanishes.

We can now appreciate why that argument is not convincing, which we brought forward against the general principle of relativity at the end of Section 18. It is certainly true that the observer in the railway carriage experiences a jerk forwards as a result of the application of the brake, and that he recognises, in this the non-uniformity of motion (retardation) of the carriage. But he is compelled by nobody to refer this jerk to a "real" acceleration (retardation) of the carriage. He might also interpret his experience thus: "My body of reference (the carriage) remains permanently at rest. With reference to it, however, there exists (during the period of application of the brakes) a gravitational field which is directed forwards and which is variable with respect to time. Under the influence of this field, the embankment together with the earth moves non-uniformly in such a manner that their original velocity in the backwards direction is continuously reduced."

TWENTY-ONE

In What Respects Are the Foundations of Classical Mechanics and of the Special Theory of Relativity Unsatisfactory?

W e have already stated several times that classical mechanics starts out from the following law: Material particles sufficiently far removed from other material particles continue to move uniformly in a straight line or continue in a state of rest. We have also repeatedly emphasised that this fundamental law can only be valid for bodies of reference K which possess certain unique states of motion, and which are in uniform translational motion relative to each other. Relative to other reference-bodies K the law is not valid.

Both in classical mechanics and in the special theory of relativity we therefore differentiate between reference-bodies K relative to which the recognised "laws of nature" can be said to hold, and reference-bodies K relative to which these laws do not hold.

But no person whose mode of thought is logical can rest satisfied with this condition of things. He asks: "How does it come that certain reference-bodies (or their states of motion) are given priority over other reference-bodies (or their states of motion)? *What is the reason for this preference?*" In order to show clearly what I mean by this question, I shall make use of a comparison.

I am standing in front of a gas range. Standing alongside of each other on the range are two pans so much alike that one may be mistaken for the other. Both are half full of water. I notice that steam is being emitted continuously from the one pan, but not from the other. I am surprised at this, even if I have never seen either a gas range or a pan before. But if I now notice a luminous something of bluish colour under the first pan but not under the other, I cease to be astonished, even if I have never before seen a gas flame. For I can only say that this bluish something will cause the emission of the steam, or at least *possibly* it may do so. If, however, I notice the bluish something in neither case, and if I observe that the one continuously emits steam whilst the other does not, then I shall remain astonished and dissatisfied until I have discovered some circumstance to which I can attribute the different behaviour of the two pans.

Analogously, I seek in vain for a real something in classical mechanics (or in the special theory of relativity) to which I can attribute the different behaviour of bodies considered with respect to the reference systems K and K^1.[1] Newton saw this objection and attempted to invalidate it, but without success. But E. Mach recognised it most clearly of all, and because of this objection he claimed that mechanics must be placed on a new basis. It can only be got rid of by means of a physics which is conformable to the general principle of relativity, since the equations of such a theory hold for every body of reference, whatever may be its state of motion.

Note

1. The objection is of importance more especially when the state of motion of the reference-body is of such a nature that it does not require any external agency for its maintenance, *e.g.* in the case when the reference-body is rotating uniformly.

TWENTY-TWO

A Few Inferences from the General Principle of Relativity

T he considerations of Section 20 show that the general principle of relativity puts us in a position to derive properties of the gravitational field in a purely theoretical manner. Let us suppose, for instance, that we know the space-time "course" for any natural process whatsoever, as regards the manner in which it takes place in the Galileian domain relative to a Galileian body of reference K. By means of purely theoretical operations (*i.e.* simply by calculation) we are then able to find how this known natural process appears, as seen from a reference-body K^1 which is accelerated relatively to K. But since a gravitational field exists with respect to this new body of reference K^1, our consideration also teaches us how the gravitational field influences the process studied.

For example, we learn that a body which is in a state of uniform rectilinear motion with respect to K (in accordance with the law of Galilei) is executing an accelerated and in general curvilinear motion with respect to the accelerated reference-body K^1 (chest). This acceleration or curvature corresponds to the influence on the moving body of the gravitational field prevailing relatively to K. It is known that a gravitational field influences the movement of bodies in this way, so that our consideration supplies us with nothing essentially new.

However, we obtain a new result of fundamental importance when we carry out the analogous consideration for a ray of light. With respect to the Galileian reference-body K, such a ray of light is transmitted rectilinearly with the velocity c. It can easily be shown that the path of the same ray of light is no longer a straight line when we consider it with reference to the accelerated chest (reference-body K^1). From this we conclude, *that, in general, rays of light are propagated curvilinearly in gravitational fields.* In two respects this result is of great importance.

In the first place, it can be compared with the reality. Although a detailed examination of the question shows that the curvature of light rays required by the general theory of relativity is only exceedingly small for the gravitational fields at our disposal, in practice, its estimated magnitude for light rays passing the sun at grazing incidence is nevertheless 1.7 seconds of arc. This ought to manifest itself in the following way. As seen from the earth, certain fixed stars appear to be in the neighbourhood of the sun, and are thus capable of observation during a total eclipse of the sun. At such times, these stars ought to appear to be displaced outwards from the sun by an amount indicated above, as compared with their apparent position in the sky when the sun is situated at another part of the heavens. The examination of the correctness or otherwise of this deduction is a problem of the greatest importance, the early solution of which is to be expected of astronomers.[1]

In the second place our result shows that, according to the general theory of relativity, the law of the constancy of the velocity of light *in vacuo,* which constitutes one of the two fundamental assumptions in the special theory of relativity and to which we have already frequently referred, cannot claim any unlimited validity. A curvature of rays of light can only take place when the velocity of propagation of light varies with position. Now we might think that as a consequence of this, the special theory of relativity and with it the whole theory of relativity would be laid in the dust. But in reality this is not the case. We can only conclude that the special theory of relativity cannot claim an unlimited domain of validity; its results hold only so long as we are able to disregard the influences of gravitational fields on the phenomena (*e.g.* of light).

Since it has often been contended by opponents of the theory of relativity that the special theory of relativity is overthrown by the general theory of relativity, it is perhaps advisable to make the facts of the case clearer by means of an appropriate comparison. Before the development of electrodynamics the laws of electrostatics were looked upon as the laws of electricity. At the present time we know that electric fields can be derived correctly from electrostatic considerations only for the case, which is never strictly realised, in which the electrical masses are quite at rest relatively to each other, and to the co-ordinate system. Should we be justified in saying that for this reason electrostatics is overthrown by the field-equations of Maxwell in electrodynamics? Not in the least. Electrostatics is contained in electrodynamics as a limiting case; the laws of the latter lead directly to those of the former for the case in which the fields are invariable with regard to time. No fairer destiny could be allotted to any physical theory, than that it should of itself point out the way to the introduction of a more comprehensive theory, in which it lives on as a limiting case.

In the example of the transmission of light just dealt with, we have seen that the general theory of relativity enables us to derive theoretically the influence of a gravitational field on the course of natural processes, the laws of which are already known when a gravitational field is absent. But the most attractive problem, to the solution of which the general theory of relativity supplies the key, concerns the investigation of the laws satisfied by the gravitational field itself. Let us consider this for a moment.

We are acquainted with space-time domains which behave (approximately) in a "Galilean" fashion under suitable choice of reference-body, i.e. domains in which gravitational fields are absent. If we now refer such a domain to a reference-body K^1 possessing any kind of motion, then relative to K^1 there exists a gravitational field which is variable with respect to space and time.[2] The character of this field will of course depend on the motion chosen for K^1. According to the general theory of relativity, the general law of the gravitational field must be satisfied for all gravitational fields obtainable in this way. Even though by no means all gravitational fields can be produced in this way, yet we may entertain the hope that the general law of gravitation will be derivable from such gravitational fields of a special kind. This hope has been realised in the most beautiful manner. But between the clear vision of this goal and its actual realisation it was necessary to surmount a serious difficulty, and as this lies deep at the root of things, I dare not withhold it from the reader. We require to extend our ideas of the space-time continuum still farther.

Notes

1. By means of the star photographs of two expeditions equipped by a Joint Committee of the Royal and Royal Astronomical Societies, the existence of the deflection of light demanded by theory was first confirmed during the solar eclipse of 29 May 1919. (Cf. Appendix III.)

2. This follows from a generalisation of the discussion in Section 20.

TWENTY THREE

Behaviour of Clocks and Measuring-Rods on a Rotating Body of Reference

Hitherto I have purposely refrained from speaking about the physical interpretation of space- and time-data in the case of the general theory of relativity. As a consequence, I am guilty of a certain slovenliness of treatment, which, as we know from the special theory of relativity, is far from being unimportant and pardonable. It is now high time that we remedy this defect; but I would mention at the outset, that this matter lays no small claims on the patience and on the power of abstraction of the reader.

We start off again from quite special cases, which we have

frequently used before. Let us consider a space-time domain in which no gravitational field exists relative to a reference-body K whose state of motion has been suitably chosen. K is then a Galileian reference-body as regards the domain considered, and the results of the special theory of relativity hold relative to K. Let us suppose the same domain referred to a second body of reference K^1, which is rotating uniformly with respect to K. In order to fix our ideas, we shall imagine K^1 to be in the form of a plane circular disc, which rotates uniformly in its own plane about its centre. An observer who is sitting eccentrically on the disc K^1 is sensible of a force which acts outwards in a radial direction, and which would be interpreted as an effect of inertia (centrifugal force) by an observer who was at rest with respect to the original reference-body K. But the observer on the disc may regard his disc as a reference-body which is "at rest"; on the basis of the general principle of relativity he is justified in doing this. The force acting on himself, and in fact on all other bodies which are at rest relative to the disc, he regards as the effect of a gravitational field. Nevertheless, the space-distribution of this gravitational field is of a kind that would not be possible on Newton's theory of gravitation.[1] But since the observer believes in the general theory of relativity, this does not disturb him; he is quite in the right when he believes that a general law of gravitation can be formulated—a law which not only explains the motion of the stars correctly, but also the field of force experienced by himself.

The observer performs experiments on his circular disc with clocks and measuring-rods. In doing so, it is his intention to arrive at exact definitions for the signification of time- and space-data with reference to the circular disc K^1, these definitions being based on his observations. What will be his experience in this enterprise?

To start with, he places one of two identically constructed clocks at the centre of the circular disc, and the other on the edge of the disc, so that they are at rest relative to it. We now ask ourselves whether both clocks go at the same rate from the standpoint of the non-rotating Galileian reference-body K. As judged from this body, the clock at the centre of the disc has no velocity, whereas the clock at the edge of the disc is in motion relative to K in consequence of the rotation. According to a result obtained in Section 12, it follows that the latter clock goes at a rate permanently slower than that of the clock at the centre of the circular disc, *i.e.* as observed from K. It is obvious that the same effect would be noted by an observer whom we will imagine sitting alongside his clock at the centre of the circular disc. Thus on our circular disc, or, to make the case more general, in every gravitational field, a clock will go more quickly or less quickly, according to the position in which the clock is situated (at rest). For this reason it is not possible to obtain a reasonable definition of time with the aid of clocks which are arranged at rest with respect to the body of reference. A similar difficulty presents itself when we attempt to apply our earlier definition of simultaneity in such a case, but I do not wish to go any farther into this question.

Moreover, at this stage the definition of the space co-ordinates also presents insurmountable difficulties. If the observer applies his standard measuring-rod (a rod which is short as compared with the radius of the disc) tangentially to the edge of the disc, then, as judged from the Galileian system, the length of this rod will be less than I, since, according to Section 12, moving bodies suffer a shortening in the direction of the motion. On the other hand, the measuring-rod will not experience a shortening in length, as judged from K, if it is applied to the disc in the direction of the radius. If, then, the observer first measures the circumference of the disc with his measuring-rod and then the diameter of the disc, on dividing the one by the other, he will not obtain as quotient the familiar number $\pi = 3.14\ldots$, but a larger number,[2] whereas, of course, for a disc which is at rest with respect to K, this operation would yield π exactly. This proves that the propositions of Euclidean geometry cannot hold exactly on the rotating disc, nor in general in a gravitational field, at least if we attribute the length I to the rod in all positions and in every orientation. Hence the idea of a straight line also loses its meaning. We are therefore not in a position to define exactly the co-ordinates x, y, z relative to the disc by means of the method used in discussing the special theory, and as long as the co-ordinates and times of events have not been defined, we cannot assign an exact meaning to the natural laws in which these occur.

Thus all our previous conclusions based on general relativity would appear to be called in question. In reality we must make a subtle detour in order to be able to apply the postulate of general relativity exactly. I shall prepare the reader for this in the following paragraphs.

Notes

1. The field disappears at the centre of the disc and increases proportionally to the distance from the centre as we proceed outwards.

2. Throughout this consideration we have to use the Galileian (non-rotating) system K as reference-body, since we may only assume the validity of the results of the special theory of relativity relative to K (relative to K[1] a gravitational field prevails).

TWENTY-FOUR

Euclidean and
Non-Euclidean Continuum

The surface of a marble table is spread out in front of me. I can get from any one point on this table to any other point by passing continuously from one point to a "neighbouring" one, and repeating this process a (large) number of times, or, in other words, by going from point to point without executing "jumps." I am sure the reader will appreciate with sufficient clearness what I mean here by "neighbouring" and by "jumps" (if he is not too pedantic). We express this property of the surface by describing the latter as a continuum.

Let us now imagine that a large number of little rods of equal length have been made, their lengths being small compared with the dimensions of the marble slab. When I say they are of equal length, I mean that one can be laid on any other without the ends overlapping. We next lay four of these little rods on the marble slab so that they constitute a quadrilateral figure (a square), the diagonals of which are equally long. To ensure the equality of the diagonals, we make use of a little testing-rod. To this square we add similar ones, each of which has one rod in common with the first. We proceed in like manner with each of these squares until finally the whole marble slab is laid out with squares. The arrangement is such, that each side of a square belongs to two squares and each corner to four squares.

It is a veritable wonder that we can carry out this business without getting into the greatest difficulties. We only need to think of the following. If at any moment three squares meet at a corner, then two sides of the fourth square are already laid, and, as a consequence, the arrangement of the remaining two sides of the square is already completely determined. But I am now no longer able to adjust the quadrilateral so that its diagonals may be equal. If they are equal of their own accord, then this is an especial favour of the marble slab and of the little rods, about which I can only be thankfully surprised. We must experience many such surprises if the construction is to be successful.

If everything has really gone smoothly, then I say that the points of the marble slab constitute a Euclidean continuum with respect to the little rod, which has been used as a "distance" (line-interval). By choosing one corner of a square as "origin" I can characterise every other corner of a square with reference to this origin by means of two numbers. I only need state how many rods I must pass over when, starting from the origin, I proceed towards the "right" and then "upwards," in order to arrive at the corner of the square under consideration. These two numbers are then the "Cartesian co-ordinates" of this corner with reference to the "Cartesian co-ordinate system" which is determined by the arrangement of little rods.

By making use of the following modification of this abstract experiment, we recognise that there must also be cases in which the experiment would be unsuccessful. We shall suppose that the rods "expand" by an amount proportional to the increase of temperature. We heat the central part of the marble slab, but not the periphery, in which case two of our little rods can still be brought into coincidence at every position on the table. But our construction of squares must necessarily come into disorder during the heating, because the little rods on the central region of the table expand, whereas those on the outer part do not.

With reference to our little rods—defined as unit lengths—the marble slab is no longer a Euclidean continuum, and we are also no longer in the position of defining Cartesian co-ordinates directly with their aid, since the above construction can no longer be carried out. But since there are other things which are not influenced in a similar manner to the little rods (or perhaps not at all) by the temperature of the table, it is possible quite naturally to maintain the point of view that the marble slab is a "Euclidean continuum." This can be done in a satisfactory manner by making a more subtle stipulation about the measurement or the comparison of lengths.

But if rods of every kind (*i.e.* of every material) were to behave *in the same way* as regards the influence of temperature when they are on the variably heated marble slab, and if we had no other means of detecting the effect of temperature than the geometrical behaviour of our rods in experiments analogous to the one described above, then our best plan would be to assign the distance *one* to two points on the slab, provided that the ends of one of our rods could be made to coincide with these two points; for how else should we define the distance without our proceeding being in the highest measure grossly arbitrary? The method of Cartesian coordinates must then be discarded, and replaced by another which does not assume the validity of Euclidean geometry for rigid bodies.[1] The reader will notice that the situation depicted here corresponds to the one brought about by the general postulate of relativity (Section 23).

Note

1. Mathematicians have been confronted with our problem in the following form. If we are given a surface (*e.g.* an ellipsoid) in Euclidean three-dimensional space, then there exists for this surface a two-dimensional geometry, just as much as for a plane surface. Gauss undertook the task of treating this two-dimensional geometry from first principles, without making use of the fact that the surface belongs to a Euclidean continuum of three dimensions. If we imagine constructions to be made with rigid rods in the surface (similar to that above with the marble slab), we should find that different laws hold for these from those resulting on the basis of Euclidean plane

geometry. The surface is not a Euclidean continuum with respect to the rods, and we cannot define Cartesian co-ordinates *in the surface*. Gauss indicated the principles according to which we can treat the geometrical relationships in the surface, and thus pointed out the way to the method of Riemman of treating multi-dimensional, non-Euclidean continuum. Thus it is that mathematicians long ago solved the formal problems to which we are led by the general postulate of relativity.

TWENTY-FIVE

Gaussian Co-ordinates

According to Gauss, this combined analytical and geometrical mode of handling the problem can be arrived at in the following way. We imagine a system of arbitrary curves (see Fig. 4) drawn on the surface of the table. These we designate as u-curves, and we indicate each of them by means of a number. The Curves u = 1, u = 2 and u = 3 are drawn in the diagram. Between the curves u = 1 and u = 2 we must imagine an infinitely large number to be drawn, all of which correspond to real numbers lying between 1 and 2. We have then a system of u-curves, and this "infinitely dense" system covers the whole surface of the table. These u-curves must not intersect each other, and through each point of the surface one

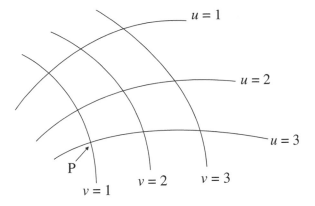

and only one curve must pass. Thus a perfectly definite value of u belongs to every point on the surface of the marble slab. In like manner we imagine a system of v-curves drawn on the surface. These satisfy the same conditions as the u-curves, they are provided with numbers in a corresponding manner, and they may likewise be of arbitrary shape. It follows that a value of u and a value of v belong to every point on the surface of the table. We call these two numbers the co-ordinates of the surface of the table (Gaussian co-ordinates). For example, the point P in the diagram has the Gaussian co-ordinates u = 3, v = 1. Two neighbouring points P and P[1] on the surface then correspond to the co-ordinates

$$P: u, v$$
$$P^1: u + du, v + dv,$$

where du and dv signify very small numbers. In a similar manner we may indicate the distance (line-interval) between P and

P^1, as measured with a little rod, by means of the very small number ds. Then according to Gauss we have

$$ds^2 = g_{11}du^2 + 2g_{12}dudv = g_{22}dv^2$$

where g_{11}, g_{12}, g_{22} are magnitudes which depend in a perfectly definite way on u and v. The magnitudes g_{11}, g_{12} and g_{22} determine the behaviour of the rods relative to the u-curves and v-curves, and thus also relative to the surface of the table. For the case in which the points of the surface considered form a Euclidean continuum with reference to the measuring-rods, but only in this case, it is possible to draw the u-curves and v-curves and to attach numbers to them, in such a manner, that we simply have:

$$ds^2 = du^2 + dv^2$$

Under these conditions, the u-curves and v-curves are straight lines in the sense of Euclidean geometry, and they are perpendicular to each other. Here the Gaussian co-ordinates are samply Cartesian ones. It is clear that Gauss co-ordinates are nothing more than an association of two sets of numbers with the points of the surface considered, of such a nature that numerical values differing very slightly from each other are associated with neighbouring points "in space."

So far, these considerations hold for a continuum of two dimensions. But the Gaussian method can be applied also to a continuum of three, four or more dimensions. If, for instance, a continuum of four dimensions be supposed available, we may represent it in the following way. With every point of the continuum, we associate arbitrarily four numbers, x_1, x_2, x_3, x_4, which are known as "co-ordinates." Adjacent points correspond to adjacent values of the coordinates. If a distance ds is associated with the adjacent points P and P^1, this distance being measurable and well defined from a physical point of view, then the following formula holds:

$$ds^2 = g_{11}dx_1^2 + 2g_{12}dx_1dx_2 \ldots g_{44}dx_4^2,$$

where the magnitudes g_{11}, etc., have values which vary with the position in the continuum. Only when the continuum is a Euclidean one is it possible to associate the co-ordinates $x_1 \ldots x_4$ with the points of the continuum so that we have simply

$$ds^2 = dx_1^2 + dx_2^2 + dx_3^2 + dx_4^2.$$

In this case relations hold in the four-dimensional continuum which are analogous to those holding in our three-dimensional measurements.

However, the Gauss treatment for ds^2 which we have given above is not always possible. It is only possible when sufficiently small regions of the continuum under consideration may be regarded as Euclidean continua. For example, this obviously holds in the case of the marble slab of the table and local variation of temperature. The temperature is practically constant for a small part of the slab, and thus the geometrical behaviour of the rods is *almost* as it ought to be according to the rules of Euclidean geometry. Hence the imperfections of the construction of squares in the previous section do not show themselves clearly until this construction is extended over a considerable portion of the surface of the table.

We can sum this up as follows: Gauss invented a method for the mathematical treatment of continua in general, in which "size-relations" ("distances" between neighbouring points) are defined. To every point of a continuum are assigned as many numbers (Gaussian co-ordinates) as the continuum has dimensions. This is done in such a way, that only one meaning can be attached to the assignment, and that numbers (Gaussian co-ordinates) which differ by an indefinitely small amount are assigned to adjacent points. The Gaussian co-ordinate system is a logical generalisation of the Cartesian co-ordinate system. It is also applicable to non-Euclidean continua, but only when, with respect to the defined "size" or "distance," small parts of the continuum under consideration behave more nearly like a Euclidean system, the smaller the part of the continuum under our notice.

TWENTY-SIX

The Space-Time Continuum
of the Special Theory of Relativity
Considered as a Euclidean
Continuum

We are now in a position to formulate more exactly the idea of Minkowski, which was only vaguely indicated in Section 17. In accordance with the special theory of relativity, certain co-ordinate systems are given preference for the description of the four-dimensional, space-time continuum. We called these "Galileian co-ordinate systems." For these systems, the four co-ordinates x, y, z, t, which determine an event—or in other words, a point of the four-dimensional continuum—are defined physically in a simple

manner, as set forth in detail in the first part of this book. For
the transition from one Galileian system to another, which is
moving uniformly with reference to the first, the equations of
the Lorentz transformation are valid. These last form the basis
for the derivation of deductions from the special theory of relativity, and in themselves they are nothing more than the expression of the universal validity of the law of transmission of light
for all Galileian systems of reference.

Minkowski found that the Lorentz transformations satisfy
the following simple conditions. Let us consider two neighbouring events, the relative position of which in the four-
dimensional continuum is given with respect to a Galileian
reference-body K by the space co-ordinate differences dx, dy, dz
and the time-difference dt. With reference to a second Galileian
system we shall suppose that the corresponding differences for
these two events are dx^1, dy^1, dz^1, dt^1. Then these magnitudes
always fulfil the condition[1]

$$dx^2 + dy^2 + dz^2 - c^2dt^2 = dx^{1\,2} + dy^{1\,2} + dz^{1\,2} - c^2dt^{1\,2}.$$

The validity of the Lorentz transformation follows from this
condition. We can express this as follows: The magnitude

$$ds^2 = dx^2 + dy^2 + dz^2 - c^2dt^2,$$

which belongs to two adjacent points of the four-dimensional
space-time continuum, has the same value for all selected

(Galileian) reference-bodies. If we replace x, y, z, $\sqrt{-1}$, by x_1, x_2, x_3, x_4, we also obtain the result that

$$ds^2 = dx_1^2 + dx_2^2 + dx_3^2 + dx_4^2.$$

is independent of the choice of the body of reference. We call the magnitude ds the "distance" apart of the two events or four-dimensional points.

Thus, if we choose as time-variable the imaginary variable $\sqrt{-1}$ instead of the real quantity t, we can regard the space-time continuum—in accordance with the special theory of relativity—as a "Euclidean" four-dimensional continuum, a result which follows from the considerations of the preceding section.

Note

1. Cf. Appendixes I and II. The relations which are derived there for the co-ordinates themselves are valid also for co-ordinate *differences,* and thus also for co-ordinate differentials (indefinitely small differences).

TWENTY-SEVEN

The Space-Time Continuum
of the General Theory of Relativity
Is Not a Euclidean Continuum

In the first part of this book we were able to make use of
space-time co-ordinates which allowed of a simple and direct physical interpretation, and which, according to Section 26, can be regarded as four-dimensional Cartesian
co-ordinates. This was possible on the basis of the law of the
constancy of the velocity of tight. But according to Section 21
the general theory of relativity cannot retain this law. On the
contrary, we arrived at the result that according to this latter
theory the velocity of light must always depend on the co-
ordinates when a gravitational field is present. In connection

with a specific illustration in Section 23, we found that the presence of a gravitational field invalidates the definition of the coordinates and the time, which led us to our objective in the special theory of relativity.

In view of the results of these considerations we are led to the conviction that, according to the general principle of relativity, the space-time continuum cannot be regarded as a Euclidean one, but that here we have the general case, corresponding to the marble slab with local variations of temperature, and with which we made acquaintance as an example of a two-dimensional continuum. Just as it was there impossible to construct a Cartesian co-ordinate system from equal rods, so here it is impossible to build up a system (reference-body) from rigid bodies and clocks, which shall be of such a nature that measuring-rods and clocks, arranged rigidly with respect to one another, shall indicate position and time directly. Such was the essence of the difficulty with which we were confronted in Section 23.

But the considerations of Sections 25 and 26 show us the way to surmount this difficulty. We refer the four-dimensional space-time continuum in an arbitrary manner to Gauss co-ordinates. We assign to every point of the continuum (event) four numbers, x_1, x_2, x_3, x_4 (co-ordinates), which have not the least direct physical significance, but only serve the purpose of numbering the points of the continuum in a definite but arbitrary manner. This arrangement does not even need to be of such a kind that we must regard x_1, x_2, x_3 as "space" co-ordinates and x_4 as a "time" co-ordinate.

The reader may think that such a description of the world

would be quite inadequate. What does it mean to assign to an event the particular co-ordinates x_1, x_2, x_3, x_4, if in themselves these co-ordinates have no significance? More careful consideration shows, however, that this anxiety is unfounded. Let us consider, for instance, a material point with any kind of motion. If this point had only a momentary existence without duration, then it would be described in space-time by a single system of values x_1, x_2, x_3, x_4. Thus its permanent existence must be characterised by an infinitely large number of such systems of values, the co-ordinate values of which are so close together as to give continuity; corresponding to the material point, we thus have a (uni-dimensional) line in the four-dimensional continuum. In the same way, any such lines in our continuum correspond to many points in motion. The only statements having regard to these points which can claim a physical existence are in reality the statements about their encounters. In our mathematical treatment, such an encounter is expressed in the fact that the two lines which represent the motions of the points in question have a particular system of co-ordinate values, x_1, x_2, x_3, x_4, in common. After mature consideration the reader will doubtless admit that in reality such encounters constitute the only actual evidence of a time-space nature with which we meet in physical statements.

When we were describing the motion of a material point relative to a body of reference, we stated nothing more than the encounters of this point with particular points of the reference-body. We can also determine the corresponding values of the time by the observation of encounters of the body with clocks, in conjunction with the observation of the encounter of the hands of clocks with particular points on the dials. It is just the same in the case of space-measurements by means of measuring-rods, as a little consideration will show.

The following statements hold generally: Every physical description resolves itself into a number of statements, each of which refers to the space-time coincidence of two events A and B. In terms of Gaussian co-ordinates, every such statement is expressed by the agreement of their four co-ordinates x_1, x_2, x_3, x_4. Thus in reality, the description of the time-space continuum by means of Gauss co-ordinates completely replaces the description with the aid of a body of reference, without suffering from the defects of the latter mode of description; it is not tied down to the Euclidean character of the continuum which has to be represented.

TWENTY-EIGHT

Exact Formulation of the General Principle of Relativity

W e are now in a position to replace the provisional formulation of the general principle of relativity given in Section 18 by an exact formulation. The form there used, "All bodies of reference K, K¹, etc., are equivalent for the description of natural phenomena (formulation of the general laws of nature), whatever may be their state of motion," cannot be maintained, because the use of rigid reference-bodies, in the sense of the method followed in the special theory of relativity, is in general not possible in space-time description. The Gauss co-ordinate system has to take the place of the body of reference. The following statement corresponds to the fundamental idea of the general principle of relativity: *"All Gaussian co-ordinate systems are essentially equivalent for the formulation of the general laws of nature."*

We can state this general principle of relativity in still another form, which renders it yet more clearly intelligible than it is when in the form of the natural extension of the special principle of relativity. According to the special theory of relativity, the equations which express the general laws of nature pass over into equations of the same form when, by making use of the Lorentz transformation, we replace the space-time variables x, y, z, t, of a (Galileian) reference-body K by the space-time variables x^1, y^1, z^1, t^1, of a new reference-body K^1. According to the general theory of relativity, on the other hand, by application of *arbitrary substitutions* of the Gauss variables x_1, x_2, x_3, x_4, the equations must pass over into equations of the same form; for every transformation (not only the Lorentz transformation) corresponds to the transition of one Gauss co-ordinate system into another.

If we desire to adhere to our "old-time" three-dimensional view of things, then we can characterise the development which is being undergone by the fundamental idea of the general theory of relativity as follows: The special theory of relativity has reference to Galileian domains, *i.e.* to those in which no gravitational field exists. In this connection a Galileian reference-body serves as body of reference, *i.e.* a rigid body the state of motion of which is so chosen that the Galileian law of the uniform rectilinear motion of "isolated" material points holds relatively to it.

Certain considerations suggest that we should refer the same Galileian domains to *non-Galileian* reference-bodies also. A gravitational field of a special kind is then present with respect to these bodies (cf. Sections 20 and 23).

In gravitational fields there are no such things as rigid bodies with Euclidean properties; thus the fictitious rigid body of reference is of no avail in the general theory of relativity. The motion of clocks is also influenced by gravitational fields, and in such a way that a physical definition of time which is made directly with the aid of clocks has by no means the same degree of plausibility as in the special theory of relativity.

For this reason non-rigid reference-bodies are used, which are as a whole not only moving in any way whatsoever, but which also suffer alterations in form *ad lib.* during their motion. Clocks, for which the law of motion is of any kind, however irregular, serve for the definition of time. We have to imagine each of these clocks fixed at a point on the non-rigid reference-body. These clocks satisfy only the one condition, that the "readings" which are observed simultaneously on adjacent clocks (in space) differ from each other by an indefinitely small amount. This non-rigid reference-body, which might appropriately be termed a "reference-mollusc," is in the main equivalent to a Gaussian four-dimensional co-ordinate system chosen arbitrarily. That which gives the "mollusc" a certain comprehensibility as compared with the Gauss co-ordinate system is the (really unjustified) formal retention of the separate existence of the space co-ordinates as opposed to the time co-ordinate. Every point on the mollusc is treated as a space-point, and every material point which is at rest relatively to it as at rest, so long as the mollusc is considered as reference-body. The general principle of relativity requires that all these molluscs can be used as reference-bodies with equal right and equal success in the formulation of the general laws of nature; the laws themselves must be quite independent of the choice of mollusc.

The great power possessed by the general principle of relativity lies in the comprehensive limitation which is imposed on the laws of nature in consequence of what we have seen above.

TWENTY-NINE

The Solution of the Problem of Gravitation on the Basis of the General Principle of Relativity

If the reader has followed all our previous considerations, he will have no further difficulty in understanding the methods leading to the solution of the problem of gravitation.

We start off on a consideration of a Galileian domain, *i.e.* a domain in which there is no gravitational field relative to the Galileian reference-body K. The behaviour of measuring-rods and clocks with reference to K is known from the special theory of relativity, likewise the behaviour of "isolated" material points; the latter move uniformly and in straight lines.

Now let us refer this domain to a random Gauss co-ordinate

system or to a "mollusc" as reference-body K^1. Then with respect to K^1 there is a gravitational field G (of a particular kind). We learn the behaviour of measuring-rods and clocks and also of freely moving material points with reference to K^1 simply by mathematical transformation. We interpret this behaviour as the behaviour of measuring-rods, docks and material points under the influence of the gravitational field G. Hereupon we introduce a hypothesis: that the influence of the gravitational field on measuring-rods, clocks and freely moving material points continues to take place according to the same laws, even in the case where the prevailing gravitational field is *not* derivable from the Galileian special care, simply by means of a transformation of co-ordinates.

The next step is to investigate the space-time behaviour of the gravitational field G, which was derived from the Galileian special case simply by transformation of the co-ordinates. This behaviour is formulated in a law, which is always valid, no matter how the reference-body (mollusc) used in the description may be chosen.

This law is not yet the *general* law of the gravitational field, since the gravitational field under consideration is of a special kind. In order to find out the general law-of-field of gravitation we still require to obtain a generalisation of the law as found above. This can be obtained without caprice, however, by taking into consideration the following demands:

a. The required generalisation must likewise satisfy the general postulate of relativity.

b. If there is any matter in the domain under consideration, only its inertial mass, and thus according to Section 15 only its energy is of importance for its effect in
exciting a field.

c. Gravitational field and matter together must satisfy
the law of the conservation of energy (and of impulse).

Finally, the general principle of relativity permits us to determine the influence of the gravitational field on the course of
all those processes which take place according to known laws
when a gravitational field is absent, *i.e.* which have already been
fitted into the frame of the special theory of relativity. In this
connection we proceed in principle according to the method
which has already been explained for measuring-rods, clocks
and freely moving material points.

The theory of gravitation derived in this way from the general postulate of relativity excels not only in its beauty; nor in
removing the defect attaching to classical mechanics which was
brought to light in Section 21; nor in interpreting the empirical
law of the equality of inertial and gravitational mass; but it has
also already explained a result of observation in astronomy,
against which classical mechanics is powerless.

If we confine the application of the theory to the case where
the gravitational fields can be regarded as being weak, and in
which all masses move with respect to the co-ordinate system
with velocities which are small compared with the velocity of
light, we then obtain as a first approximation the Newtonian
theory. Thus the latter theory is obtained here without any

particular assumption, whereas Newton had to introduce the hypothesis that the force of attraction between mutually attracting material points is inversely proportional to the square of the distance between them. If we increase the accuracy of the calculation, deviations from the theory of Newton make their appearance, practically all of which must nevertheless escape the test of observation owing to their smallness.

We must draw attention here to one of these deviations. According to Newton's theory, a planet moves round the sun in an ellipse, which would permanently maintain its position with respect to the fixed stars, if we could disregard the motion of the fixed stars themselves and the action of the other planets under consideration. Thus, if we correct the observed motion of the planets for these two influences, and if Newton's theory be strictly correct, we ought to obtain for the orbit of the planet an ellipse, which is fixed with reference to the fixed stars. This deduction, which can be tested with great accuracy, has been confirmed for all the planets save one, with the precision that is capable of being obtained by the delicacy of observation attainable at the present time. The sole exception is Mercury, the planet which lies nearest the sun. Since the time of Leverrier, it has been known that the ellipse corresponding to the orbit of Mercury, after it has been corrected for the influences mentioned above, is not stationary with respect to the fixed stars, but that it rotates exceedingly slowly in the plane of the orbit and in the sense of the orbital motion. The value obtained for this rotary movement of the orbital ellipse was 43 seconds of arc per century, an amount ensured to be correct to within a few seconds of arc. This effect can be explained by means of classical mechanics only on the assumption of hypotheses which have little probability, and which were devised solely for this purpose.

On the basis of the general theory of relativity, it is found that the ellipse of every planet round the sun must necessarily rotate in the manner indicated above; that for all the planets, with the exception of Mercury, this rotation is too small to be detected with the delicacy of observation possible at the present time; but that in the case of Mercury it must amount to 43 seconds of arc per century, a result which is strictly in agreement with observation.

Apart from this one, it has hitherto been possible to make only two deductions from the theory which admit of being tested by observation, to wit, the curvature of light rays by the gravitational field of the sun,[1] and a displacement of the spectral lines of light reaching us from large stars, as compared with the corresponding lines for light produced in an analogous manner terrestrially (*i.e.* by the same kind of atom).[2] These two deductions from the theory have both been confirmed.

Notes

1. First observed by Eddington and others in 1919. (Cf. Appendix III.)
2. Established by Adams in 1924. (Cf. Appendix III.)

PART THREE

Considerations on the Universe as a Whole

THIRTY

Cosmological Difficulties
of Newton's Theory

Apart from the difficulty discussed in Section 21, there is a second fundamental difficulty attending classical celestial mechanics, which, to the best of my knowledge, was first discussed in detail by the astronomer Seeliger. If we ponder over the question as to how the universe, considered as a whole, is to be regarded, the first answer that suggests itself to us is surely this: As regards space (and time) the universe is infinite. There are stars everywhere, so that the density of matter, although very variable in detail, is nevertheless on the average everywhere the same. In other words: However far we might travel through space, we should find everywhere an attenuated swarm of fixed stars of approximately the same kind and density.

This view is not in harmony with the theory of Newton. The latter theory rather requires that the universe should have a kind of centre in which the density of the stars is a maximum, and that as we proceed outwards from this centre the group-density of the stars should diminish, until finally, at great distances, it is succeeded by an infinite region of emptiness. The stellar universe ought to be a finite island in the infinite ocean of space.[1]

This conception is in itself not very satisfactory. It is still less satisfactory because it leads to the result that the light emitted by the stars and also individual stars of the stellar system are perpetually passing out into infinite space, never to return, and without ever again coming into interaction with other objects of nature. Such a finite material universe would be destined to become gradually but systematically impoverished.

In order to escape this dilemma, Seeliger suggested a modification of Newton's law, in which he assumes that for great distances the force of attraction between two masses diminishes more rapidly than would result from the inverse square law. In this way it is possible for the mean density of matter to be constant everywhere, even to infinity, without infinitely large gravitational fields being produced. We thus free ourselves from the distasteful conception that the material universe ought to possess something of the nature of a centre. Of course we purchase our emancipation from the fundamental difficulties mentioned, at the cost of a modification and complication of Newton's law which has neither empirical nor theoretical foundation. We can imagine innumerable laws which would serve the same purpose, without our being able to state a reason why one of them is to be preferred to the others; for any one of these laws would be founded just as little on more general theoretical principles as is the law of Newton.

Note

1. *Proof*—According to the theory of Newton, the number of "lines of force" which come from infinity and terminate in a mass m is proportional to the mass m. If, on the average, the mass density p_0 is constant throughout the universe, then a sphere of volume V will enclose the average man p_0V. Thus the number of lines of force passing through the surface F of the sphere into its interior is proportional to p_0V. For unit area of the surface of the sphere the number of lines of force which enters the sphere is thus proportional to p_0V/F or to p_0R. Hence the intensity of the field at the surface would ultimately become infinite with increasing radius R of the sphere, which is impossible.

THIRTY-ONE

The Possibility of a "Finite" and Yet "Unbounded" Universe

But speculations on the structure of the universe also move in quite another direction. The development of non-Euclidean geometry led to the recognition of the fact, that we can cast doubt on the infiniteness of our space without coming into conflict with the laws of thought or with experience (Riemann, Helmholtz). These questions have already been treated in detail and with unsurpassable lucidity by Helmholtz and Poincaré, whereas I can only touch on them briefly here.

In the first place, we imagine an existence in two-dimensional space. Flat beings with flat implements, and in particular flat rigid measuring-rods, are free to move in a *plane*. For them nothing exists outside of this plane: that which they observe to happen to themselves and to their flat "things" is the all-inclusive reality of their plane. In particular, the constructions of plane Euclidean geometry can be carried out by means of the rods, *e.g.* the lattice construction, considered in Section 24. In contrast to ours, the universe of these beings is two-dimensional; but, like ours, it extends to infinity. In their universe there is room for an infinite number of identical squares made up of rods, *i.e.* its volume (surface) is infinite. If these beings say their universe is "plane," there is sense in the statement, because they mean that they can perform the constructions of plane Euclidean geometry with their rods. In this connection the individual rods always represent the same distance, independently of their position.

Let us consider now a second two-dimensional existence, but this time on a spherical surface instead of on a plane. The flat beings with their measuring-rods and other objects fit exactly on this surface and they are unable to leave it. Their whole universe of observation extends exclusively over the surface of the sphere. Are these beings able to regard the geometry of their universe as being plane geometry and their rods withal as the realisation of "distance"? They cannot do this. For if they attempt to realise a straight line, they will obtain a curve, which we "three-dimensional beings" designate as a great circle, *i.e.* a self-contained line of definite finite length, which can be measured up by means of a measuring-rod. Similarly, this universe has a finite area that can be compared with the area of a square constructed with rods. The great charm resulting from this consideration lies in the recognition of the fact that the universe of these beings is finite and yet has no limits.

But the spherical-surface beings do not need to go on a world-tour in order to perceive that they are not living in a Euclidean universe. They can convince themselves of this on every part of their "world," provided they do not use too small a piece of it. Starting from a point, they draw "straight lines" (arcs of circles as judged in three-dimensional space) of equal length in all directions. They will call the line joining the free ends of these lines a "circle." For a plane surface, the ratio of the circumference of a circle to its diameter, both lengths being measured with the same rod, is, according to Euclidean geometry of the plane, equal to a constant value π, which is independent of the diameter of the circle. On their spherical surface our flat beings would find for this ratio the value

$$\pi \frac{\sin\left(\dfrac{r}{R}\right)}{\left(\dfrac{r}{R}\right)}$$

i.e. a smaller value than π, the difference being the more considerable, the greater is the radius of the circle in comparison with the radius R of the "world-sphere." By means of this relation the spherical beings can determine the radius of their universe ("world"), even when only a relatively small part of their world-sphere is available for their measurements. But if this part is very small indeed, they will no longer be able to demonstrate that they are on a spherical "world" and not on a Euclidean plane, for a small part of a spherical surface differs only slightly from a piece of a plane of the same size.

Thus if the spherical surface beings are living on a planet of which the solar system occupies only a negligibly small part of the spherical universe, they have no means of determining whether they are living in a finite or in an infinite universe, because the "piece of universe" to which they have access is in both cases practically plane, or Euclidean. It follows directly from this discussion, that for our sphere-beings the circumference of a circle first increases with the radius until the "circumference of the universe" is reached, and that it thenceforward gradually decreases to zero for still further increasing values of the radius. During this process the area of the circle continues to increase more and more, until finally it becomes equal to the total area of the whole "world-sphere."

Perhaps the reader will wonder why we have placed our "beings" on a sphere rather than on another closed surface. But this choice has its justification in the fact that, of all closed surfaces, the sphere is unique in possessing the property that all points on it are equivalent. I admit that the ratio of the circumference c of a circle to its radius r depends on r, but for a given value of r it is the same for all points of the "world-sphere"; in other words, the "world-sphere" is a "surface of constant curvature."

To this two-dimensional sphere-universe there is a three-dimensional analogy, namely, the three-dimensional spherical space which was discovered by Riemann. Its points are likewise all equivalent. It possesses a finite volume, which is determined by its "radius" $(2\pi^2R^3)$. Is it possible to imagine a spherical space? To imagine a space means nothing else than that we imagine an epitome of our "space" experience, *i.e.* of experience that we can have in the movement of "rigid" bodies. In this sense we can imagine a spherical space.

Suppose we draw lines or stretch strings in all directions from a point, and mark off from each of these the distance r with a measuring-rod. All the free end-points of these lengths lie on a spherical surface. We can specially measure up the area (F) of this surface by means of a square made up of measuring-rods. If the universe is Euclidean, then $F = 4\pi R^2$; if it is spherical, then F is always less than $4\pi R^2$. With increasing values of r, F increases from zero up to a maximum value which is determined by the "world-radius," but for still further increasing values of r, the area gradually diminishes to zero. At first, the straight lines which radiate from the starting point diverge farther and farther from one another, but later they approach each other, and finally they run together again at a "counter-point" to the starting point. Under such conditions they have traversed the whole spherical space. It is easily seen that the three-dimensional spherical space is quite analogous to the two-dimensional spherical surface. It is finite (*i.e.* of finite volume), and has no bounds.

It may be mentioned that there is yet another kind of curved space: "elliptical space." It can be regarded as a curved space in which the two "counter-points" are identical (indistinguishable from each other). An elliptical universe can thus be considered to some extent as a curved universe possessing central symmetry.

It follows from what has been said, that closed spaces without limits are conceivable. From amongst these, the spherical space (and the elliptical) excels in its simplicity, since all points on it are equivalent. As a result of this discussion, a most interesting question arises for astronomers and physicists, and that is whether the universe in which we live is infinite, or whether it is finite in the manner of the spherical universe. Our experience is far from being sufficient to enable us to answer this question. But the general theory of relativity permits of our answering it with a moderate degree of certainty, and in this connection the difficulty mentioned in Section 30 finds its solution.

THIRTY-TWO

The Structure of Space
According to the General
Theory of Relativity

According to the general theory of relativity, the geometrical properties of space are not independent, but they are determined by matter. Thus we can draw conclusions about the geometrical structure of the universe only if we base our considerations on the state of the matter as being something that is known. We know from experience that, for a suitably chosen co-ordinate system, the velocities of the stars are small as compared with the velocity of transmission of light. We can thus as a rough approximation arrive at a conclusion as to the nature of the universe as a whole, if we treat the matter as being at rest.

We already know from our previous discussion that the behaviour of measuring-rods and clocks is influenced by gravitational fields, *i.e.* by the distribution of matter. This in itself is sufficient to exclude the possibility of the exact validity of Euclidean geometry in our universe. But it is conceivable that our universe differs only slightly from a Euclidean one, and this notion seems all the more probable, since calculations show that the metrics of surrounding space is influenced only to an exceedingly small extent by masses even of the magnitude of our sun. We might imagine that, as regards geometry, our universe behaves analogously to a surface which is irregularly curved in its individual parts, but which nowhere departs appreciably from a plane: something like the rippled surface of a lake. Such a universe might fittingly be called a quasi-Euclidean universe. As regards its space it would be infinite. But calculation shows that in a quasi-Euclidean universe the average density of matter would necessarily be *nil*. Thus such a universe could not be inhabited by matter everywhere; it would present to us that unsatisfactory picture which we portrayed in Section 30.

If we are to have in the universe an average density of matter which differs from zero, however small may be that difference, then the universe cannot be quasi-Euclidean. On the contrary, the results of calculation indicate that if matter be distributed uniformly, the universe would necessarily be spherical (or elliptical). Since in reality the detailed distribution of matter is not uniform, the real universe will deviate in individual parts from the spherical, *i.e.* the universe will be quasi-spherical. But it will be necessarily finite. In fact, the theory supplies us with a simple connection[1] between the space-expanse of the universe and the average density of matter in it.

Note

1. For the radius R of the universe we obtain the equation

$$R^2 = \frac{2}{kp}$$

The use of the C.G.S. system in this equation gives $2/k = 1.08 . 10^{27}$; p is the average density of the matter and k is a constant connected with the Newtonian constant of gravitation.

Simple Derivation of the Lorentz Transformation

(Supplementary to Section 11)

For the relative orientation of the co-ordinate systems indicated in Fig. 2, the x-axes of both systems permanently coincide. In the present case we can divide the problem into parts by considering first only events which are localised on the x-axis. Any such event is represented with respect to the co-ordinate system K by the abscissa x and the time t, and with respect to the system K^1 by the abscissa x′ and the time t′. We require to find x′ and t′ when x and t are given.

APPENDIX B

A light-signal, which is proceeding along the positive axis of x, is transmitted according to the equation

$$x = ct$$

or

$$x - ct = 0 \tag{1}$$

Since the same light-signal has to be transmitted relative to K^1 with the velocity c, the propagation relative to the system K^1 will be represented by the analogous formula

$$x' - ct' = 0 \tag{2}$$

Those space-time points (events) which satisfy (x) must also satisfy (2). Obviously this will be the case when the relation

$$(x' - ct') = \lambda (x - ct) \tag{3}$$

is fulfilled in general, where λ indicates a constant; for, according to (3), the disappearance of $(x - ct)$ involves the disappearance of $(x' - ct')$.

If we apply quite similar considerations to light rays which are being transmitted along the negative x-axis, we obtain the condition

$$(x' + ct') = \mu(x + ct) \tag{4}$$

By adding (or subtracting) equations (3) and (4), and introducing for convenience the constants a and b in place of the constants λ and μ, where

$$a = \frac{\lambda + \mu}{2}$$

and

$$b = \frac{\lambda - \mu}{2}$$

we obtain the equations

$$\left. \begin{array}{l} x^1 = ax - bct \\ ct^1 = act - bx \end{array} \right\} \qquad (5)$$

We should thus have the solution of our problem, if the constants a and b were known. These result from the following discussion.

For the origin of K^1 we have permanently $x' = 0$, and hence according to the first of the equations (5)

$$x = \frac{bc}{a} t$$

If we call v the velocity with which the origin of K^1 is moving relative to K, we then have

$$v = \frac{bc}{a} \tag{6}$$

The same value v can be obtained from equations (5), if we calculate the velocity of another point of K^1 relative to K, or the velocity (directed towards the negative x-axis) of a point of K with respect to K'. In short, we can designate v as the relative velocity of the two systems.

Furthermore, the principle of relativity teaches us that, as judged from K, the length of a unit measuring-rod which is at rest with reference to K^1 must be exactly the same as the length, as judged from K', of a unit measuring-rod which is at rest relative to K. In order to see how the points of the x-axis appear as viewed from K, we only require to take a "snapshot" of K^1 from K; this means that we have to insert a particular value of t (time of K), *e.g.* t = 0. For this value of t we then obtain from the first of the equations (5)

$$x' = ax$$

Two points of the x'-axis which are separated by the distance $\Delta x' = I$ when measured in the K^1 system are thus separated in our instantaneous photograph by the distance

$$\Delta x = \frac{I}{a} \tag{7}$$

But if the snapshot be taken from K′(t′ = 0), and if we eliminate t from the equations (5), taking into account the expression (6), we obtain

$$x^1 = a\left(I - \frac{v^2}{c^2}\right)x$$

From this we conclude that two points on the x-axis separated by the distance I (relative to K) will be represented on our snapshot by the distance

$$\Delta x^1 = a\left(I - \frac{v^2}{c^2}\right) \qquad (7a)$$

But from what has been said, the two snapshots must be identical; hence Δx in (7) must be equal to $\Delta x'$ in (7a), so that we obtain

$$a = \frac{I}{I - \frac{v^2}{c^2}} \qquad (7b)$$

The equations (6) and (7b) determine the constants a and b. By inserting the values of these constants in (5), we obtain the first and the fourth of the equations given in Section 11.

$$
\left.
\begin{aligned}
x' &= \frac{x - vt}{\sqrt{I - \dfrac{v^2}{c^2}}} \\[2em]
t' &= \frac{t - \dfrac{v}{c^2} x}{\sqrt{I - \dfrac{v^2}{c^2}}}
\end{aligned}
\right\}
\tag{8}
$$

Thus we have obtained the Lorentz transformation for events on the x-axis. It satisfies the condition

$$
x'^2 - c^2 t'^2 = x^2 - c^2 t^2
\tag{8a}
$$

The extension of this result, to include events which take place outside the x-axis, is obtained by retaining equations (8) and supplementing them by the relations

$$
\left.
\begin{aligned}
y' &= y \\
z' &= z
\end{aligned}
\right\}
\tag{9}
$$

In this way we satisfy the postulate of the constancy of the velocity of light *in vacuo* for rays of light of arbitrary direction, both for the system K and for the system K′. This may be shown in the following manner.

We suppose a light-signal sent out from the origin of K at the time t = 0. It will be propagated according to the equation

$$
r = \sqrt{x^2 + y^2 + z^2} = ct
$$

or, if we square this equation, according to the equation

$$x^2 + y^2 + z^2 = c^2t^2 = 0 \qquad (10)$$

It is required by the law of propagation of light, in conjunction with the postulate of relativity, that the transmission of the signal in question should take place—as judged from K^1—in accordance with the corresponding formula

$$r' = ct'$$

or,

$$x'^2 + y'^2 + z'^2 - c^2t'^2 = 0 \qquad (10a)$$

In order that equation (10a) may be a consequence of equation (10), we must have

$$x'^2 + y'^2 + z'^2 - c^2t'^2 = \sigma \left(x^2 + y^2 + z^2 - c^2t^2 \right) \qquad (11)$$

Since equation (8a) must hold for points on the x-axis, we thus have $\sigma = I$. It is easily seen that the Lorentz transformation really satisfies equation (11) for $\sigma = I$; for (11) is a consequence of (8a) and (9), and hence also of (8) and (9). We have thus derived the Lorentz transformation.

The Lorentz transformation represented by (8) and (9) still requires to be generalised. Obviously it is immaterial whether the axes of K^1 be chosen so that they are spatially parallel to those of K. It is also not essential that the velocity of translation of K¹ with respect to K should be in the direction of the x-axis. A simple consideration shows that we are able to construct the Lorentz transformation in this general sense from two kinds of transformations, *viz.* from Lorentz transformations in the special sense and from purely spatial transformations, which corresponds to the replacement of the rectangular co-ordinate system by a new system with its axes pointing in other directions.

Mathematically, we can characterise the generalised Lorentz transformation thus:

It expresses x', y', x', t', in terms of linear homogeneous functions of x, y, x, t, of such a kind that the relation

$$x'^2 + y'^2 + z'^2 - c^2 t'^2 = x^2 + y^2 + z^2 - c^2 t^2 \qquad (11a)$$

is satisfied identically. That is to say: If we substitute their expressions in x, y, x, t, in place of x', y', x', t', on the left-hand side, then the left-hand side of (11a) agrees with the right-hand side.

Minkowski's Four-Dimensional Space ("World")

(Supplementary to Section 17)

W e can characterise the Lorentz transformation still more simply if we introduce the imaginary $\sqrt{-I \cdot ct}$ in place of t, as time-variable. If, in accordance with this, we insert

$$x_1 = x$$
$$x_2 = y$$
$$x_3 = z$$
$$x_4 = \sqrt{-I \cdot ct}$$

and similarly for the accented system K^1, then the condition which is identically satisfied by the transformation can be expressed thus:

$$x_1'^2 + x_2'^2 + x_3'^2 + x_4'^2 = x_1^2 + x_2^2 + x_3^2 + x_4^2 \tag{12}$$

That is, by the aforementioned choice of "co-ordinates," (11a) [see the end of Appendix II] is transformed into this equation.

We see from (12) that the imaginary time co-ordinate x_4 enters into the condition of transformation in exactly the same way as the space co-ordinates x_1, x_2, x_3. It is due to this fact that, according to the theory of relativity, the "time" x_4 enters into natural laws in the same form as the space co-ordinates x_1, x_2, x_3.

A four-dimensional continuum described by the "co-ordinates" x_1, x_2, x_3, x_4, was called "world" by Minkowski, who also termed a point-event a "world-point." From a "happening" in three-dimensional space, physics becomes, as it were, an "existence" in the four-dimensional "world."

This four-dimensional "world" bears a close similarity to the three-dimensional "space" of (Euclidean) analytical geometry. If we introduce into the latter a new Cartesian co-ordinate system (x_1', x_2', x_3') with the same origin, then x_1', x_2', x_3', are linear homogeneous functions of x_1, x_2, x_3 which identically satisfy the equation

$$x_1'^2 + x_2'^2 + x_3'^2 = x_1^2 + x_2^2 + x_3^2$$

The analogy with (12) is a complete one. We can regard Minkowski's "world" in a formal manner as a four-dimensional Euclidean space (with an imaginary time co-ordinate); the Lorentz transformation corresponds to a "rotation" of the co-ordinate system in the four-dimensional "world."

The Experimental Confirmation of the General Theory of Relativity

From a systematic theoretical point of view, we may imagine the process of evolution of an empirical science to be a continuous process of induction. Theories are evolved and are expressed in short compass as statements of a large number of individual observations in the form of empirical laws, from which the general laws can be ascertained by comparison. Regarded in this way, the development of a science bears some resemblance to the compilation of a classified catalogue. It is, as it were, a purely empirical enterprise.

But this point of view by no means embraces the whole of the actual process; for it slurs over the important part played by intuition and deductive thought in the development of an exact science. As soon as a science has emerged from its initial stages, theoretical advances are no longer achieved merely by a process of arrangement. Guided by empirical data, the investigator rather develops a system of thought which, in general, is built up logically from a small number of fundamental assumptions, the so-called axioms. We call such a system of thought a *theory*. The theory finds the justification for its existence in the fact that it correlates a large number of single observations, and it is just here that the "truth" of the theory lies.

Corresponding to the same complex of empirical data, there may be several theories, which differ from one another to a considerable extent. But as regards the deductions from the theories which are capable of being tested, the agreement between the theories may be so complete that it becomes difficult to find any deductions in which the two theories differ from each other. As an example, a case of general interest is available in the province of biology, in the Darwinian theory of the development of species by selection in the struggle for existence, and in the theory of development which is based on the hypothesis of the hereditary transmission of acquired characters.

We have another instance of far-reaching agreement between the deductions from two theories in Newtonian mechanics on the one hand, and the general theory of relativity on the other. This agreement goes so far, that up to the present we have been able to find only a few deductions from the general theory of relativity which are capable of investigation, and to which the physics of pre-relativity days does not also lead, and this despite the profound difference in the fundamental assumptions of the two theories. In what follows, we shall again consider these important deductions, and we shall also discuss the empirical evidence appertaining to them which has hitherto been obtained.

A. Motion of the Perihelion
of Mercury

According to Newtonian mechanics and Newton's law of gravitation, a planet which is revolving round the sun would describe an ellipse round the latter, or, more correctly, round the common centre of gravity of the sun and the planet. In such a system, the sun, or the common centre of gravity, lies in one of the foci of the orbital ellipse in such a manner that, in the course of a planet-year, the distance sun-planet grows from a minimum to a maximum, and then decreases again to a minimum. If instead of Newton's law we insert a somewhat different law of attraction into the calculation, we find that, according to this new law, the motion would still take place in such a manner that the distance sun-planet exhibits periodic variations; but in this case the angle described by the line joining sun and planet during such a period (from perihelion—closest proximity to the sun—to perihelion) would differ from 360°. The line of the orbit would not then be a closed one but in the course of time it would fill up an annular part of the orbital plane, *viz.* between the circle of least and the circle of greatest distance of the planet from the sun.

According also to the general theory of relativity, which differs of course from the theory of Newton, a small variation from the Newton-Kepler motion of a planet in its orbit should take place, and in such a way, that the angle described by the radius sun-planet between one perihelion and the next should exceed that corresponding to one complete revolution by an amount given by

$$+ \frac{24\pi^3 a^2}{T^2 c^2 (I - e^2)}$$

(Note: One complete revolution corresponds to the angle 2π in the absolute angular measure customary in physics, and the above expression gives the amount by which the radius sun-planet exceeds this angle during the interval between one perihelion and the next.) In this expression a represents the major semi-axis of the ellipse, e its eccentricity, c the velocity of light, and T the period of revolution of the planet. Our result may also be stated as follows: According to the general theory of relativity, the major axis of the ellipse rotates round the sun in the same sense as the orbital motion of the planet. Theory requires that this rotation should amount to 43 seconds of arc per century for the planet Mercury, but for the other planets of our solar system its magnitude should be so small that it would necessarily escape detection.[1]

In point of fact, astronomers have found that the theory of Newton does not suffice to calculate the observed motion of Mercury with an exactness corresponding to that of the delicacy of observation attainable at the present time. After taking account of all the disturbing influences exerted on Mercury by the remaining planets, it was found (Leverrier, 1859; and Newcomb, 1895) that an unexplained perihelial movement of the orbit of Mercury remained over, the amount of which does not differ sensibly from the above mentioned + 43 seconds of arc per century. The uncertainty of the empirical result amounts to a few seconds only.

B. Deflection of Light by a Gravitational Field

In Section 22 it has been already mentioned that according to the general theory of relativity, a ray of light will experience a curvature of its path when passing through a gravitational field, this curvature being similar to that experienced by the path of a body which is projected through a gravitational field. As a result of this theory, we should expect that a ray of light which is passing close to a heavenly body would be deviated towards the latter. For a ray of light which passes the sun at a distance of Δ sun-radii from its centre, the angle of deflection (a) should amount to

$$a = \frac{1.7 \text{ seconds of arc}}{\Delta}$$

It may be added that, according to the theory, half of this deflection is produced by the Newtonian field of attraction of the sun, and the other half by the geometrical modification ("curvature") of space caused by the sun.

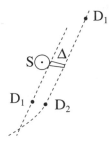

This result admits of an experimental test by means of the photographic registration of stars during a total eclipse of the sun. The only reason why we must wait for a total eclipse is because at every other time the atmosphere is so strongly illuminated by the light from the sun that the stars situated near the sun's disc are invisible. The predicted effect can be seen clearly from the accompanying diagram. If the sun (S) were not present, a star which is practically infinitely distant would be seen in the direction D_1, as observed from the earth. But as a consequence of the deflection of light from the star by the sun, the star will be seen in the direction D_2, *i.e.* at a somewhat greater distance from the centre of the sun than corresponds to its real position.

In practice, the question is tested in the following way. The stars in the neighbourhood of the sun are photographed during a solar eclipse. In addition, a second photograph of the same stars is taken when the sun is situated at another position in the sky, *i.e.* a few months earlier or later. As compared with the standard photograph, the positions of the stars on the eclipse-photograph ought to appear displaced radially outwards (away from the centre of the sun) by an amount corresponding to the angle a.

We are indebted to the [British] Royal Society and to the Royal Astronomical Society for the investigation of this important deduction. Undaunted by the [first world] war and by difficulties of both a material and a psychological nature aroused by the war, these societies equipped two expeditions—to Sobral (Brazil), and to the island of Principe (West Africa)—and sent several of Britain's most celebrated astronomers (Eddington, Cottingham, Crommelin, Davidson), in order to obtain photographs of the solar eclipse of 29 May 1919. The relative discrepancies to be expected between the stellar photographs obtained during the eclipse and the comparison photographs amounted to a few hundredths of a millimetre only. Thus great accuracy was necessary in making the adjustments required for the taking of the photographs, and in their subsequent measurement.

The results of the measurements confirmed the theory in a thoroughly satisfactory manner. The rectangular components of the observed and of the calculated deviations of the stars (in seconds of arc) are set forth in the following table of results:

Number of the Star		First Co-ordinate		Second Co-ordinate	
		Observed	Calculated	Observed	Calculated
11	. .	−0.19	−0.22	+0.16	+0.02
5	. .	+0.29	+0.31	−0.46	−0.43
4	. .	+0.11	+0.10	+0.83	+0.74
3	. .	+0.20	+0.12	+1.00	+0.87
6	. .	+0.10	+0.04	+0.57	+0.40
10	. .	−0.08	+0.09	+0.35	+0.32
2	. .	+0.95	+0.85	−0.27	−0.09

C. *Displacement of Spectral Lines Towards the Red*

In Section 23 it has been shown that in a system K^1 which is in rotation with regard to a Galileian system K, clocks of identical construction, and which are considered at rest with respect to the rotating reference-body, go at rates which are dependent on the positions of the clocks. We shall now examine this dependence quantitatively. A clock, which is situated at a distance r from the centre of the disc, has a velocity relative to K which is given by

$$V = wr$$

where w represents the angular velocity of rotation of the disc K^1 with respect to K. If v_0, represents the number of ticks of the clock per unit time ("rate" of the clock) relative to K when the clock is at rest, then the "rate" of the clock (v) when it is moving relative to K with a velocity V, but at rest with respect to the disc, will, in accordance with Section 12, be given by

$$v = v_2\sqrt{I - \frac{v^2}{c^2}}$$

or with sufficient accuracy by

$$v = v_0\left(I - \frac{1}{2}\frac{v^2}{c^2}\right)$$

This expression may also be stated in the following form:

$$v = v_0 \left(I - \frac{1}{c^2} \frac{w^2 r^2}{2} \right)$$

If we represent the difference of potential of the centrifugal force between the position of the clock and the centre of the disc by φ, *i.e.* the work, considered negatively, which must be performed on the unit of mass against the centrifugal force in order to transport it from the position of the clock on the rotating disc to the centre of the disc, then we have

$$\phi = \frac{w^2 r^2}{2}$$

From this it follows that

$$v = v_0 \left(I + \frac{\phi}{c^2} \right)$$

In the first place, we see from this expression that two clocks of identical construction will go at different rates when situated at different distances from the centre of the disc. This result is also valid from the standpoint of an observer who is rotating with the disc.

Now, as judged from the disc, the latter is in a gravitational field of potential φ, hence the result we have obtained will hold quite generally for gravitational fields. Furthermore, we can regard an atom which is emitting spectral lines as a clock, so that the following statement will hold:

An atom absorbs or emits light of a frequency which is dependent on the potential of the gravitational field in which it is situated.

The frequency of an atom situated on the surface of a heavenly body will be somewhat less than the frequency of an atom of the same element which is situated in free space (or on the surface of a smaller celestial body).

Now $\varphi = -K(M/r)$, where K is Newton's constant of gravitation, and M is the mass of the heavenly body. Thus a displacement towards the red ought to take place for spectral lines produced at the surface of stars as compared with the spectral lines of the same element produced at the surface of the earth, the amount of this displacement being

$$\frac{v_0 - v}{v_0} = \frac{K}{c^2}\frac{M}{r}$$

For the sun, the displacement towards the red predicted by theory amounts to about two millionths of the wave-length. A trustworthy calculation is not possible in the case of the stars, because in general neither the mass M nor the radius r is known.

It is an open question whether or not this effect exists, and at the present time (1920) astronomers are working with great zeal towards the solution. Owing to the smallness of the effect in the case of the sun, it is difficult to form an opinion as to its existence. Whereas Grebe and Bachem (Bonn), as a result of their own measurements and those of Evershed and Schwarzschild on the cyanogen bands, have placed the existence of the effect almost beyond doubt, while other investigators, particularly St. John, have been led to the opposite opinion in consequence of their measurements.

Mean displacements of lines towards the less refrangible end of the spectrum are certainly revealed by statistical investigations of the fixed stars; but up to the present the examination of the available data does not allow of any definite decision being arrived at, as to whether or not these displacements are to be referred in reality to the effect of gravitation. The results of observation have been collected together, and discussed in detail from the standpoint of the question which has been engaging our attention here, in a paper by E. Freundlich entitled "Zur Prüfung der allgemeinen Relativitäts-Theorie" (*Die Naturwissenschaften,* 1919, no. 35, p. 520: Julius Springer, Berlin).

At all events, a definite decision will be reached during the next few years. If the displacement of spectral lines towards the red by the gravitational potential does not exist, then the general theory of relativity will be untenable. On the other hand, if the cause of the displacement of spectral lines be definitely traced to the gravitational potential, then the study of this displacement will furnish us with important information as to the mass of the heavenly bodies.[2]

Notes

1. Especially since the next planet Venus has an orbit that is almost an exact circle, which makes it more difficult to locate the perihelion with precision.

2. The displacement of spectral lines towards the red end of the spectrum was definitely established by Adams in 1924, by observations on the dense companion of Sirius, for which the effect is about thirty times greater than for the Sun.

The Structure of Space According to the General Theory of Relativity

(Supplementary to Section 32)

Since the publication of the first edition of this little book, our knowledge about the structure of space in the large ("cosmological problem") has had an important development, which ought to be mentioned even in a popular presentation of the subject.

My original considerations on the subject were based on two hypotheses:

THE STRUCTURE OF SPACE

1. There exists an average density of matter in the whole of space which is everywhere the same and different from zero.
2. The magnitude ("radius") of space is independent of time.

Both these hypotheses proved to be consistent, according to the general theory of relativity, but only after a hypothetical term was added to the field equations, a term which was not required by the theory as such nor did it seem natural from a theoretical point of view ("cosmological term of the field equations").

Hypothesis (2) appeared unavoidable to me at the time, since I thought that one would get into bottomless speculations if one departed from it.

However, already in the twenties, the Russian mathematician Friedman showed that a different hypothesis was natural from a purely theoretical point of view. He realised that it was possible to preserve hypothesis (1) without introducing the less natural cosmological term into the field equations of gravitation, if one was ready to drop hypothesis (2). Namely, the original field equations admit a solution in which the "world radius" depends on time (expanding space). In that sense one can say, according to Friedman, that the theory demands an expansion of space.

A few years later Hubble showed, by a special investigation of the extra-galactic nebulae ("milky ways"), that the spectral lines emitted showed a red shift which increased regularly with the distance of the nebulae. This can be interpreted in regard to our present knowledge only in the sense of Doppler's principle, as an expansive motion of the system of stars in the large—as required, according to Friedman, by the field equations of gravitation. Hubble's discovery can, therefore, be considered to some extent as a confirmation of the theory.

There does arise, however, a strange difficulty. The interpretation of the galactic line-shift discovered by Hubble as an expansion (which can hardly be doubted from a theoretical point of view), leads to an origin of this expansion which lies "only" about 10^9 years ago, while physical astronomy makes it appear likely that the development of individual stars and systems of stars takes considerably longer. It is in no way known how this incongruity is to be overcome.

I further want to remark that the theory of expanding space, together with the empirical data of astronomy, permit no decision to be reached about the finite or infinite character of (three-dimensional) space, while the original "static" hypothesis of space yielded the closure (finiteness) of space.